Indoor Radon - Sources, Transport Mechanisms and Influencing Parameters

Authored by Christian Di Carlo, Andrea Maiorana and Francesco Bochicchio

Published in London, United Kingdom

Indoor Radon - Sources, Transport Mechanisms and Influencing Parameters
http://dx.doi.org/10.5772/intechopen.112088
Authored by Christian Di Carlo, Andrea Maiorana and Francesco Bochicchio

Contributors
Christian Di Carlo, Andrea Maiorana and Francesco Bochicchio

First published in London, United Kingdom, 2023 by IntechOpen
IntechOpen is the global imprint of INTECHOPEN LIMITED, registered in England and Wales, registration number: 11086078, 5 Princes Gate Court, London, SW7 2QJ, United Kingdom

British Library Cataloguing-in-Publication Data
A catalogue record for this book is available from the British Library

Additional hard and PDF copies can be obtained from orders@intechopen.com

Indoor Radon - Sources, Transport Mechanisms and Influencing Parameters
Authored by Christian Di Carlo, Andrea Maiorana and Francesco Bochicchio
p. cm.
Print ISBN 978-0-85014-180-1
Online ISBN 978-0-85014-181-8
eBook (PDF) ISBN 978-0-85014-182-5

Meet the authors

Christian Di Carlo obtained a Ph.D. in Energy and Environment from Sapienza University of Rome (Italy) in 2021. He has been working as a researcher at the National Center for Radiation Protection and Computational Physics of the Italian National Institute of Health since 2020. He has published about twenty papers in indexed scientific journals and contributed to the proceedings of numerous national and international conferences. He contributed to the European Atlas of Natural Radiation and to the last edition of the World Health Organization's (WHO) Guidelines for Drinking Water Quality (GDWQ). He has been nominated for the United Nations Scientific Committee on the Effects of Atomic Radiation (UNSCEAR) and currently collaborates with the WHO on radioactivity in drinking water and the response to nuclear and radiological emergencies.

Andrea Maiorana is a Ph.D. student in Energy and Environment at the National Center for Radiation Protection and Computational Physics of the Italian National Institute of Health. He graduated in 2021 from Sapienza – University of Rome (Italy) with a Master's Degree in Nuclear Engineering. His research is concerned with natural radioactivity sources with a focus on radon-related issues.

Francesco Bochicchio has been working at the Italian National Institute of Health since 1987, first as a researcher and then as a senior researcher, director of research, and director of the National Center of Radiation Protection and Computational Physics. He is currently the director of the National Center for Radiation Protection and Computational Physics. He has collaborated with several international organizations, such as the European Commission (EC)/Euratom, International Atomic Energy Agency (IAEA), United Nations Scientific Committee on the Effects of Atomic Radiation (UNSCEAR), and the World Health Organization (WHO). He also contributed, as a national representative or WHO consultant, to the development of the European and the International Basic Safety Standards on protection from ionizing radiation.

Contents

Preface IX

Acknowledgements XI

Chapter 1
Indoor Radon: Sources, Transport Mechanisms and Influencing Parameters
by Christian Di Carlo, Andrea Maiorana and Francesco Bochicchio

Preface

Radon is a naturally occurring radioactive gas produced by the spontaneous decay of radium, which, in turn, is formed from the decay of uranium. Uranium is a constituent of the earth's crust and it is present everywhere. However, its concentration is variable over a wide range depending on chemical and geological processes occurring throughout the years. Radon atoms rapidly decay in other solid elements, commonly referred to as radon progeny or radon decay products.

Radon atoms are generated in the soil and in materials obtained from the manipulation of the earth's crust. The difference in terms of radon concentration and air pressure between inside and outside the generating matrix results in the exhalation of radon atoms into the surrounding environment. Radon is rapidly diluted in outdoor air, whereas it may accumulate in closed spaces.

The breathing process is the main route of radon intake into the human body. Most of the radon inhaled is exhaled before decaying. On the contrary, radon decay products, which are attached to the air particulate and consequently inhaled together with radon gas, deposit on the surfaces of the respiratory tract where they irradiate the nearby cells. The radiation energy released may damage the target cells, potentially resulting in lung cancer. Despite that most of the damages are due to radon progeny decay, the term "radon exposure" is generally preferred instead of "radon progeny exposure."

The population's exposure to indoor radon is a leading cause of lung cancer. The results of many epidemiological studies have shown an increase in lung cancer risk with radon exposure as well as a synergic effect between radon and cigarette smoking. An increase in lung cancer risk has been evidenced even at low radon concentrations.

Several national and international regulations have been targeting the public health risk of radon exposure. Although soil represents the main source of indoor radon, it is important to consider any and all sources. The relative contribution of each source depends on specific circumstances, mainly the chemical and physical characteristics of the generating matrix and the building specifics.

The generation of radon atoms and their transport indoors are complex processes due to the large number of influencing parameters and phenomena to be considered simultaneously. This book systematically addresses the current state of the art of radon generation and transport. It presents and develops mathematical models to describe the physical processes involved. It also examines the impact of influencing parameters, both physical and chemical, using the latest research.

Christian Di Carlo, Andrea Maiorana and Francesco Bochicchio
Italian National Institute of Health,
National Center for Radiation Protection and Computational Physics,
Rome, Italy

Acknowledgements

The authors would like to express their gratitude to the scientists involved in the activities related to the subject addressed in this book. Furthermore, a special thanks goes to Professor Romolo Remetti who transmitted his passion and inspired this work.

Chapter

Indoor Radon: Sources, Transport Mechanisms and Influencing Parameters

Christian Di Carlo, Andrea Maiorana and Francesco Bochicchio

Abstract

Population exposure to indoor radon has been proven to increase the risk of lung cancer, and it is considered a leading cause after tobacco smoking. Due to the relatively low outdoor activity concentration, most of the exposure to radon occurs indoors. Radon is generated mostly by the rocks that contain radium-226 either in the soil or in the building materials. Once generated, it enters the buildings directly, due to the activity concentration gradient, or indirectly via a radon carrier. The magnitude and the relative contribution of the entry patterns depend mainly on the building characteristics, the geology, and the living habits of occupants.

Keywords: indoor air, radon, radon sources, radon transport, natural radioactivity

1. Introduction

Radon is a naturally occurring radioactive noble gas (Z = 86). It is gaseous in almost all the environmental conditions found on Earth; at atmospheric pressure, the boiling point is −61,8°C. In nature, there are three radon isotopes: radon-222, radon-220, and radon-219, generally referred to as radon, thoron, and actinon, respectively. The most abundant is the ^{222}Rn, which has a decay half-life of 3.823 days [1, 2]. This isotope is produced by the alpha decay of ^{226}Ra and belongs to the natural decay chains of ^{238}U, one of the constituents of Earth's undisturbed crust.

The worldwide average per capita annual effective dose to the public is estimated to be about 3 mSv y^{-1}. Approximately half of the effective dose comes from radon and thoron (**Figure 1**) [3]. The United Nations Scientific Committee on the Effects of Atomic Radiation (UNSCEAR) reported 1.15 and 0.1 mSv y^{-1} due to inhalation of radon and thoron, respectively. The half-life of radon is long relative to the human breathing process, so almost all the energy released in the respiratory tract comes from the short-lived decay products. Nevertheless, "exposure to radon" is commonly used to indicate the effective dose due to radon short-lived decay products.

Radon is a leading cause of lung cancer [4, 5], established to be a Group 1 and Group A human carcinogen, according to the classification used by the International Agency for Research on Cancer [6]. Strong evidences of the association existing between the exposure to radon at home and lung cancer have been reported [7–9].

IntechOpen

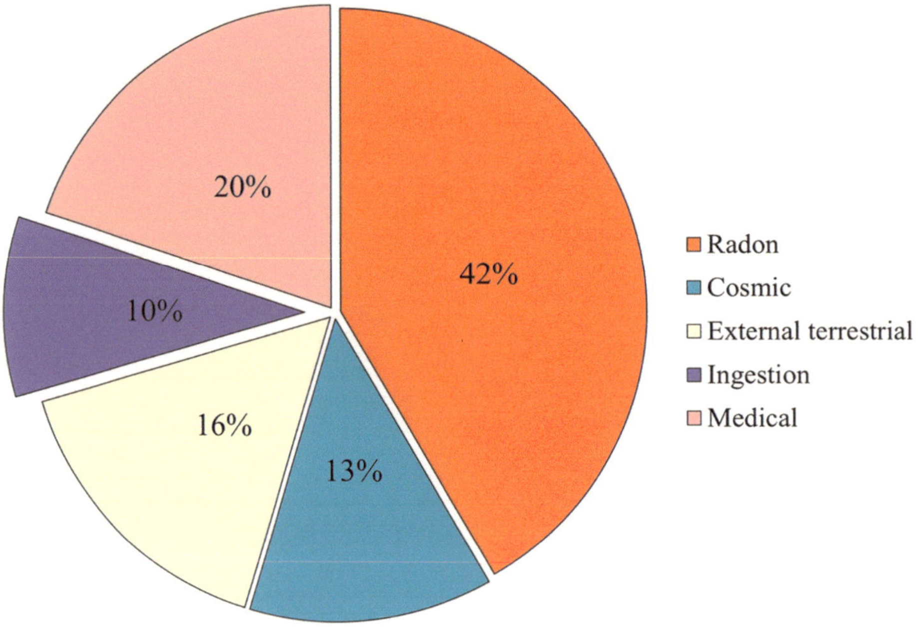

Figure 1.
Most recent assessments of the worldwide average exposures (in mSv y^{-1}*) of the public [3]. "Radon" accounts for all the sources giving dose to the public through inhalation, i.e., radon, thoron, and other radionuclides of uranium and thorium series.*

Several studies have been carried out to investigate the correlation between radon exposure and health effects other than lung cancers. Ecological studies suggested a positive correlation between radon exposure and adult acute leukemia and childhood leukemia [6], and a cohort study in Switzerland, based on estimated radon levels and not on direct measurements, showed an association between radon exposure and skin cancer mortality [10].

Due to the relatively low outdoor radon activity concentration, most of the exposure to the radon progeny occurs indoors. The typical value of radon activity concentration outdoors at ground level was reported to be 10 Bq m^{-3} [11]. More recent results [12, 13] have confirmed such estimates.

The indoor activity concentration of radon decay products widely varies [14]. This is mainly due to the spatial variability of indoor radon activity concentration, but the variability of the equilibrium factor contributes as well[1]. Several studies have dealt with temporal (e.g., [15–17]) and spatial (e.g., [15, 18–20]) variability of indoor radon activity concentration. According to the World Health Organization (WHO) [5], indoor radon activity concentration varies with two main factors: the typology of building construction and the ventilation characteristics. In the recent past, the impact of ventilation on the activity concentration indoors was studied with a systematic approach, for example by Collignan and Powaga [21], Chao, Tung [22]; however, the major source of spatial and temporal variability of radon activity concentration has been observed to be the different entry rates from the various sources.

[1] The ratio of the actual short-lived radon decay product activity concentration to the theoretical activity concentration if all the decay products were in equilibrium with the parent radon-222.

Radon is generated mostly by the rocks that contain radium-226 either in the soil or in the building materials. Once generated, it enters the buildings directly, due to an activity concentration gradient, or indirectly via a radon carrier (mainly air and domestic water and gas). If the carrier is the air, the radon flux is driven by a pressure difference between the inside and outside of the building.

The absolute and relative contribution of each of these entry patterns depends on the specific circumstances of a building [23]: mainly its construction characteristics (including building materials, construction typology, foundation typology, floor level, and building age) [24] and the geology of the underlying soil [25], in addition to the hydrology and morphology of surroundings. Many models have been proposed and adopted to estimate the contributions of each source to the indoor radon activity concentration (e.g., [11, 26, 27]). However, the results obtained cannot be generalized because they strongly depend on the site-specific model input adopted.

The direct ingress from soil usually prevails on the other sources [28, 29], especially for detached houses [30]; in particular, the main mechanism for radon entry is the pressure-driven flow of soil gas through floor cracks [28].

UNSCEAR discusses three distinct actual models to assess the contributions to the indoor radon activity concentration: a wooden and a masonry house modeled by Arvela [27] and a third masonry house model [11, 26, 31]. Results are summarized in **Table 1**.

Indoor radon activity concentration results from a balance between radon inflows from the sources and outflows due to the radioactive decay and the ventilation. In practice, either the inflows or the outflows are strongly time-dependent, so the steady state is never reached in current scenarios. The environmental conditions and the occupants' behavior are the main reasons why inflows and outflows change throughout the time [32]. The resulting radon activity concentration is characterized by variability components of different time scales.

Two periodic components mainly exist: a short-term component on a daily scale due to the diurnal cycle of temperature and pressure and the reduction of ventilation

Radon entry rate		Wooden house (Arvela, 1995)	Concrete house (Arvela, 1995)	Masonry house (UNSCEAR, 1993)	Masonry house (UNSCEAR, 2000)
Building materials	**Diffusion**	**3%**	**18%**	**21%**	**20%**
Subjacent soil	Diffusion	6%	4%	15%	25%
	Advection[1]	86%	73%	41%	35%
Outdoor air	Infiltration	4%	4%	20%	18%
Water	Degassing	1%	1%	2%	2%
Natural gas	Degassing			1%	<1%
Total (Bq m^{-3} h^{-1})		70	90	49	56

[1]In the following, as in literature, "advection" and "convection" are used equivalently regarding radon migration. The same applies to "advective" and "convective".
Results refer to a model masonry building with a volume of 250 m^3, a surface area of 450 m^2, and an air exchange rate of 1 h^{-1} [26].

Table 1.
Summary of relative contributions to indoor radon accumulation coming from all the sources.

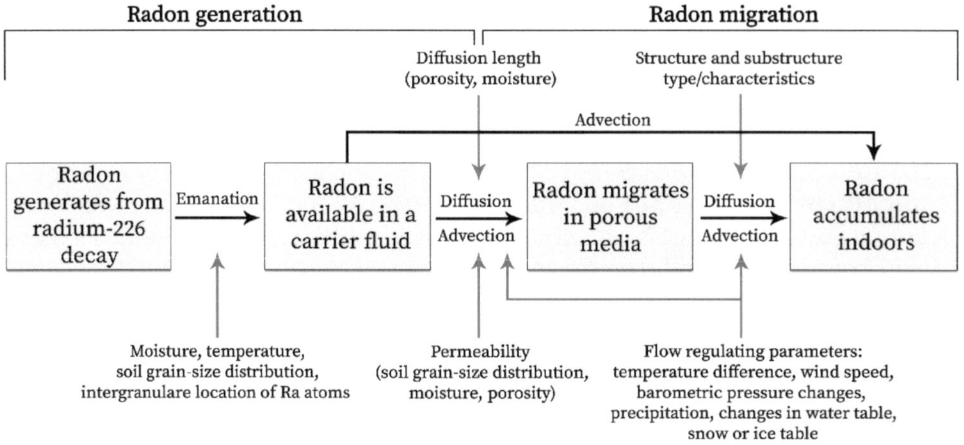

Figure 2.
Schematic representation (modified on that from Nazaroff and Nero [29]) of the different steps taking radon from the radioactive creation to the ingrowth indoors. The horizontal arrows represent the physical phenomena that allow radon to start its migration, moving from solid matrix. Chemical and physical parameters affecting the migration phenomena are placed close to vertical arrows.

in the night and a long-term component on an annual scale due to the seasonality of environmental parameters and living habits [33, 34]. Indoor radon levels are generally higher during the day than in the night and in the cold season than in the warm ones. However, some specific exceptions are documented to exist [35–37]. Furthermore, short- and medium-term aperiodical variability components affect the radon levels as well and are mainly associated to episodic or persisting meteorological phenomena.

The radon activity concentration indoors results from two subsequent phases (**Figure 2**): (i) radon is generated from Ra-226 atoms and becomes available for migration, then (ii) it migrates down an activity concentration and/or pressure gradient.

2. Radon generation

Porous material volume is generally distinguished in two main portions: (i) the solid fraction consisting mainly of mineral grains of different sizes and of a small amount of organic matter and (ii) the void fraction, which can be filled with liquid (generally water) and gas (generally similar to air in terms of composition) (**Figure 3**).

Moving from this general concept, porosity is defined as the ratio of the volume of voids to the total volume [38].

$$\varepsilon = \frac{V_v}{V} \tag{1}$$

$$V_v = V_w + V_a \tag{2}$$

$$V = V_w + V_a + V_s \tag{3}$$

where V_v stands for the volume of the void fraction (water, V_w, and air, V_a) and V for the total volume (made of void, V_v, and solid grains, V_s).

Interstitial water-filled volume, V_w Solid grains, V_s Void gas-filled volume, V_a

Figure 3.
Structure of typical soil partitioning into three volumes: Solid, water-filled, and gas-filled.

Moisture content is defined as the ratio of void volume filled with water to the total volume:

$$m = \frac{V_w}{V} \tag{4}$$

According to the latter formulation, being $V_w \leq V_v$, the moisture content cannot exceed the porosity. When $V_w = V_v$, the degree of saturation is 100%, which means the entire pore volume is filled with water.

The saturation degree, S, is defined as:

$$S = \frac{V_w}{V_v} = \frac{V_w}{V_a + V_w} \tag{5}$$

Radon atoms in porous materials can be regarded as present in both portions, in solid ("bound radon") and pore one ("free radon"). They always generate in solid grains (sometimes referred to as crystalline lattice), and only a certain fraction of them finally results to be in pores depending on the recoil origin, length, and direction. This phenomenon is generally referred to as emanation.

The radioactive decay of radium-226[2] is the original point of the radon generation process. Radon can escape from the soil grain in which it is generated by alpha particle recoil or diffusion. Because of the very low diffusion coefficient of radon in the solid grain [39], the recoil always prevails. Radon atoms from the alpha decay of radium-226 have an initial kinetic energy of 86 keV.

$$^{226}_{88}\text{Ra} \overset{\alpha\ decay}{\rightarrow} {}^{222}_{86}\text{Rn} + {}^{4}_{2}\text{He}^{2+} + Q_{value} \tag{6}$$

$$Q_{value} = 4870.62\ \text{keV} \tag{7}$$

[2] The nuclear data for radioactive decay are taken from Chisté and Bé [1], and Martin [2].

$m_{0,{}^{226}_{88}\text{Ra}} = 226.025406$ uma

$m_{0,{}^{222}_{86}\text{Rn}} = 222.017574$ uma

$m_{0,{}^{4}_{2}\text{He}} = 4.002603$ uma

$$E_{^{222}_{86}\mathrm{Rn}} = 86.21\ \mathrm{keV} \tag{8}$$

where Q_{value} is the energy released by the nuclear decay reaction. The distance traveled by radon atoms can be computed from their energy. This length depends on the density and composition of the material. Tanner [40] reported the recoil distance to range between 0.02 and 0.07 μm in common materials and to be about 0.1 and 63 μm in water and air, respectively (slightly lower values for air and water were found by Sakoda and Ishimori [41]).

The emanation process is simplified as the sum of a direct and an indirect component [40, 41]. Referring to **Figure 4**, the direct component consists of radon atoms that leave the grain due to recoil and enter air- (path A in **Figure 4**) or water-filled pores (B in **Figure 4**). The indirect component is constituted by radon atoms that are firstly trapped in adjacent grains and then diffuse from the pocket created by its recoil passage into pore (E) or radon atoms recoiled into an inner grain pore (F) then diffused out in an outer one.

The emanated fraction does not include the radon atoms whose recoil finishes in a grain, either the original (C and G) or an adjacent one (D), because they generally decay before escaping the grain by diffusion as those which remain are adsorbed on the surface of inner pores.

Summarizing, the radon emanation coefficient (also indicated in the literature as radon emanation fraction or power) accounts for the number of radon atoms released into air- and water-filled pores compared to the number of radon atoms generated in grains.

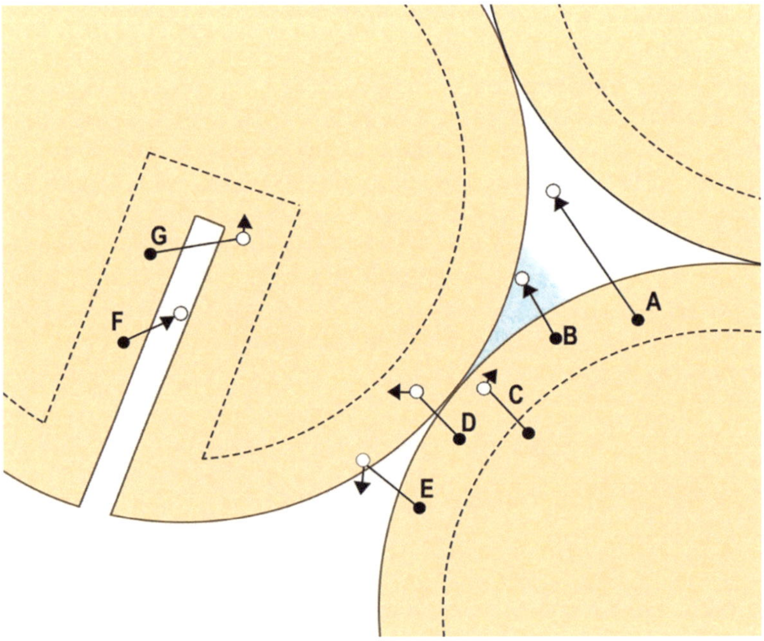

Figure 4.
Schematic diagram of radon emanation process rearranged from Sakoda and Ishimori [41]. The full and empty dots stand for the starting and the ending point of the recoil, respectively.

Material	Emanation coefficient (%)	Range (%)
Mineral	3%	
Rock	13%	2.1–32%
Soil	20%	0.14–80%
Building materials		0.1–58%
Mill tailings (mostly uranium)	17%	
Fly ash	3%	

Table 2.
Typical radon emanation coefficient [41, 42].

The radon emanation production rate per unit pore volume assesses the activity released in pores per unit time; if the radon fraction in water-filled pores is assumed negligible, it can be expressed as [29]:

$$G_v = \frac{1}{\varepsilon} I \rho \eta \lambda_{\mathrm{Rn}} \tag{9}$$

Where:

G_v is the radon emanation production rate, or the radon production rate, per unit pore volume (Bq s^{-1} m^{-3}),

I is the activity concentration of radium per unit mass of the material (Bq kg^{-1}),

ρ is the density of the material (kg m^{-3}),

η is the emanation coefficient,

λ_{Rn} is the radon decay constant (s^{-1}).

Typical values of radon emanation coefficient for minerals, rocks, soil, mill tailings, and fly ashes are reported in **Table 2**. Radon emanation coefficient is material-specific because it depends on the material's inner properties, that is, mineralogical and chemical composition, and physical parameters, as well as on environmental conditions, that is, humidity and temperature (see §2.2.2 and §2.3.2).

According to **Table 2**, soil typically has a higher radon emanation coefficient than building materials. This may be explained by the position of radium-226 atoms in soil grains that is found to be closer to the grain surfaces than in building materials [41] (see §3.1 and §4.1).

2.1 Soil

Soil is defined as the uppermost layer of Earth's crust. It is mainly composed of rocks weathered and accumulated on the surface. The soil composition strongly varies [42], but it is generally made of about 45% of minerals, 25% of water, 25% of air, and 5% of organic matter. The solid rock beneath the soil is called "bedrock" [43].

Typically, the activity concentration of natural radionuclide in soil is lower than in the bedrock: this is justified by the capability of mobilization and retention of the soil and the bedrock, mainly due to presence of water and organic matter that both may affect the radionuclides mobility [44]. Some exceptions due to site-specific conditions are documented in the literature [45].

Referring to the soil as a source of indoor radon means considering the radon atoms produced by the radioactive decay of radium-226 in the earthen materials underlying the building structure.

2.1.1 Radium-226 activity concentration

Radon and radium-226 activity concentration typically increase with the depth in soil [42]. However, different radon trends are documented for specific situations, for example, ground water tables [46].

Indoor radon is contributed to only by a limited volume of soil beneath the building, sometimes referred to as disturbed soil [32]. The radon activity concentration in disturbed soil is not constant in time because it is affected by environmental parameters [47], for example, atmospheric pressure [48, 49], outdoor temperature, rainfall, and snow/ice as ground covers [42]. Seasonal variations of the soil radon activity concentration are documented in the literature (e.g., [50, 51]).

Radon activity concentrations measured in soils usually range between 5 and 100 kBq m^{-3} [42]. Although many attempts have been made to map the geogenic radon, that is, the radon activity concentration in soil, no European map exists nowadays. The geogenic radon potential has been recently introduced to account for the susceptibility of a location to geogenic radon [52].

The radium-226 activity concentration is used to evaluate the soil strength as the radon source. Over 1500 soil samples from different geological contexts (mainly from Nepal and France) were measured by Perrier, Girault [53]: a mean value of 7.5 Bq kg^{-1} was found, with 90% of values between 1.4 and 28 Bq kg^{-1}. A non-systematic summary of radium activity concentration experimentally measured in soil over the years is reported in **Table 3**.

Soil typology	Radium-226		Reference
	Typical activity concentration (Bq kg^{-1})	Range (Bq kg^{-1})	
Surface soils	37	2.4—430	[54]
Sand and silt	—	5—25	[54]
Clay	—	20—120	[54]
Moraine	—	20—80	[54]
Soils with alum shale	—	100—1000	[54]
Dried homogeneous sand	—	3.55—3,81	[54]
Dense glacial till	—	77.5—216	[54]
Esker sand	75		[54]
Silt	52	38—63[a]	[54]
Sand	54	31—61[a]	[54]
Gravel	81	53—100[a]	[54]
Till	126	56—95[a]	[54]
Shale bearing soil	—	720—1760	[54]
Alluvial deposits	35	6.5—93	[55]
Continental deposits	31	6.5—55	[55]
Marine deposits	27	6.5—55	[55]
Flysch	32	6.5—113	[55]
Platform Carbonate	36	6.5—60	[55]
Volcanic soil	—	55—617	[55]

Soil typology	Radium-226		Reference
	Typical activity concentration (Bq kg^{-1})	**Range (Bq kg^{-1})**	
Lava	150	106—194[b]	[56]
Tuff and tuffite	146	44—248 [b]	[56]
Sedimentary rocks	109	20—198[b]	[56]
Soils derived from dolomite	—	67—98	[45]

[a]*50% of measured values found in this range.*
[b]*68% of measured values found in this range.*

Table 3.
Typical radon emanation coefficient [41, 42].

The Joint Research Centre (JRC) has recently made available the European map of uranium activity concentration in soil [57], created using approximately 5000 data from topsoil samples (collected at 0–20 cm depth) collected in national databases (Belgium, Czech Republic and Estonia) and two main European databases: (i) the Geochemical Atlas of Europe and (ii) the Geochemical Mapping of Agricultural and Grazing Land soil in Europe.

The different chemical properties (mainly mobility and retention) cause the disequilibrium to exist in the decay chain of uranium-238 [58]. Equilibrium conditions between uranium-238 and radium-226 are rarely verified [44, 59], but some examples of quasi-equilibrium are documented [56, 60].

2.1.2 Radon emanation and influencing parameters

Several experimental measurements of the emanation coefficient have been carried out through the years (e.g., [61, 62]). A summary of results from the late 90's is reported in **Table 4**. Measurements on soil samples from nine different countries (Austria, Denmark, Finland, Japan, Jordan, Slovenia, Spain, Sweden, USA) were collected in 2011 by Sakoda and Ishimori [41]. The same authors reported the emanation fractions of minerals, rocks, mill tailings, and fly ashes.

Soil description	Emanation coefficient[a]	Moisture content	Reference
Soil	0.03–0.55	Unknown	[29]
Soil[b]	0.25	Dried, 105°C, 24 h	[29]
Soil[c]	0.68	20% of dry weight	[29]
	0.09	Dried, 200°C, 90 h	[29]
Soil	0.41	Air-dry[d]	[29]
Soil	0.22–0.32	13–20% of dry weight	[29]
Soil	0.30–0.46	4% of dry weight	[29]
Alluvium soil	0.17	Unknown	[41]
Eluvium	0.20–0.38	Unknown	[41]
Volcanic	0.49	Unknown	[41]
Lava fields	0.02	Unknown	[29]
Volcanic ash soils (deep agricultural soils)	0.70	Unknown	[29]

Soil description	Emanation coefficient[a]	Moisture content	Reference
Thin organic soils	0.55	Unknown	[29]
Clay	0.17–0.40	Unknown	[29, 54]
	0.03–0.2	Dry	[41]
Stony clay	0.02–0.03	Dry	[41]
Diluvial clay (glacial deposit)	0.12–0.27	0.15—0.27 wt.	[41]
Silt	0.11–0.24[e]	Unknown	[54]
Silty loam	0.18–0.40	Unknown	[29]
Silt disaggregated loam	0.15–0.29	Moist	[41]
Sand	0.016–0.276	Unknown	[29, 41, 54]
	0.243	Saturated	[29]
	0.13–0.28	0–1 wt.	[41]
	0.002–0.08	Dry	[41]
Esker sand	0.2	Unknown	[41, 54]
Sandy loam	0.10–0.36	Unknown	[29]
Sandy disaggregated loam	0.13–0.21	Moist	[41]
Gravely sandy loam	0.38	Unknown	[41, 54]
Gravel	0.15–0.23[e]	Unknown	[54]
	0.097–0.125	Dry	[41]
Diluvial sand, silt and gravel	0.09–0.32	0—0.24 wt.	[41]
Moraine clay	0.11–0.39	0.03—0.22 wt.	[41]
Moraine sand	0.09–0.25	0.04—0.13 wt.	[41]
Rock *(crushed)*	0.005–0.40	Unknown	[29]
Limestone (soil from)	0.05–0.42	0—0.4	[41]
Calcareous soil	0.21–0.53	Unknown	[41]
Limestone and chalk	0.02–0.17	0—0.16 wt.	[41]
Granitic	0.37–0.46	Unknown	[41]
Granitic gneiss (soil from)	0.05–0.4	0—0.2	[41]
Shale (soil from)	0.02–0.32	0—0.4	[41]
Glauconitic soil	0.03	Unknown	[41]
Lignitic	0.36	Unknown	[41]
Residual (Baux)	0.14–0.30	Unknown	[41]
Residual (Calc.)	0.52–0.54	Unknown	[41]
Alum-shale bearing soil	0.09–0.69	0—0.33	[41]
Uranium Ore	0.06–0.26	Unknown	[29]
	0.014–0.07	Dried, 110°C	[29]
Uranium Ore *(crushed)*	0.055–0.55	Moist, saturated	[29]
	0.023–0.36	Vacuum-dried	[29]
Tailings	0.30	Saturated	[29]
	0.07	Dried, 110°C	[29]

Soil description	Emanation coefficient[a]	Moisture content	Reference
Dense glacial till	0.15–0.24	Unknown	[54]
Till	0.12–0.25[e]	Unknown	[54]

[a]Ranges or arithmetic means according to how data are presented in the reference.[b]Sample sieved through 20 μm mesh.[c]Sample of six giving highest reading.[d]Exposed to air laboratory for several days.[e]50% of values were found in this range.

Table 4.
Measurements of ^{222}Rn emanation coefficients, data are taken from Nazaroff and Nero [29], Font [54] and Sakoda and Ishimori [41].

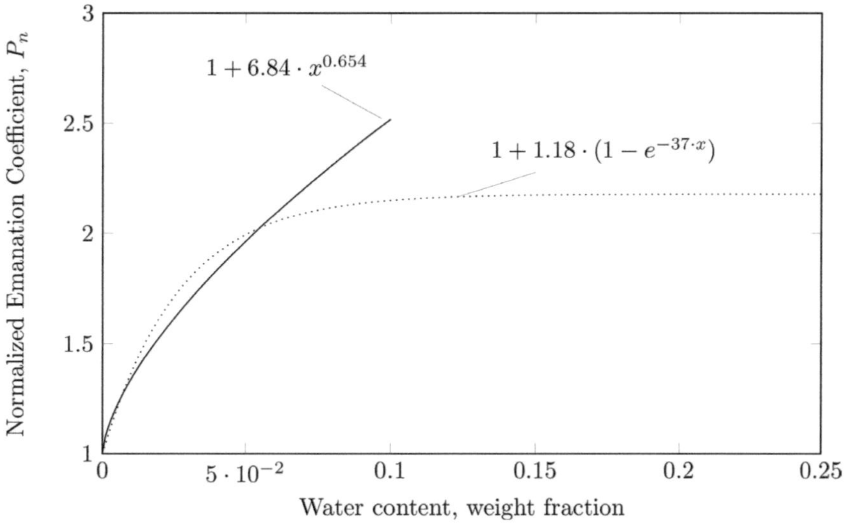

Figure 5.
Emanation coefficient as a function of the water content of soil. Values are normalized to the emanation power with no water. The fitting functions are determined by [62] for soil samples. For small water contents (<10%), a different function (plain line) better describes the dependency than the one (dotted line) for the whole range of humidity (0–30%). Same results were found by [61].

The typical value of the emanation coefficient for soils is around 0.25 [41], but it widely varies over the entire range.

The main parameters influencing the radon emanation coefficient are radium distribution in grains, grain size and shape, moisture content, temperature, and pore size [41, 63, 64].

The lower radon range in water than in air largely explains the impact of moisture content on the increase of the emanation coefficient (**Table 4**). Strong and Levins [65] reported an increase of about 370% when the moisture content increased from 0.2 to 5.7% (weight percentage). The radon emanation coefficient has been observed to increase up to the saturation reached at a specific moisture content (**Figure 5**). According to the results found by Phong Thu, Van Thang [61] for five different samples sieved to obtain different grain size (from <0.1 mm to >0.5 mm), the smaller the particle size, the higher is the moisture content necessary to achieve saturation (4–16%). The increase of pore-gap length[3] with grain size explains this behavior: a larger

[3] As defined by Sakoda, Ishimori [41], pore-gap is the gap between a radium-bearing grain and a neighboring grain with which the recoil radon can collide.

grain size increases the probability for radon atoms to lose their energy while traveling in pores, thus reducing the amount of water required to effectively stop the radon atoms during the recoil. Some authors found the moisture content effect on radon emanation to be independent from grain size [66].

Font and Baixeras [32] introduced the effective emanation coefficient and a formulation to consider the overall effects of moisture content and grain size on the radon emanation coefficient:

$$f' = fH = f_{max}[0.2 + 0.8(1 - e^{-qm})]H \tag{10}$$

Where:

H is the fraction of radon atoms emanated into the pore volume that reaches the gas volume,

m is the water saturation fraction (dimensionless),

f_{max} is the emanation fraction at saturation conditions, assumed to be maximum,

and $q = 18 + 2\log(d)$ considering the effect of the mean soil grain diameter, d (μm).

The temperature has a great influence on the radon emanation coefficient. Zhang and Zhang [64] provided a systematic review (**Figure 6**) of findings about the behavior of emanation coefficient over the temperature (0—800°C) of the soil sample (clay, red mud and coal shales). Radon emanation was observed to decrease with increasing temperature because the mean pore size increases as well with the

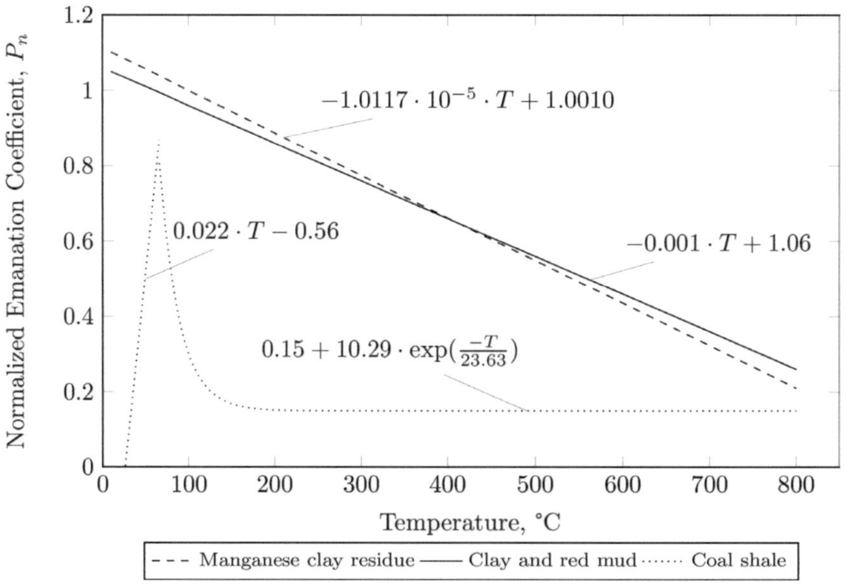

Figure 6.
Comparison of emanation coefficient behavior for three different types of soil. The fitting functions have been collected by Zhang et al. [64]. The normalization is different for the three functions: The first is normalized to 100°C (25%), the second to 200°C (18% for clay and 7% for red mud) or 400°C (26% for red mud) depending on the specific soil type, and the third to 65°C (30%). The different shapes are due to the temperature gradient that gets established inside the sample: Relatively large for manganese clay and lower for coal shale.

temperature [67, 68]. Some soils were shown to experience an initial slight increase with increasing temperature mainly due to the evaporation of inner water.

Furthermore, at the typical environmental temperature (5–40°C), small temperature reductions lead to the decrease of radon atoms available in pores for further migration even if the emanation coefficient is substantially unaffected: this happens due to the physical adsorption (usually referred to as Van der Waal adsorption) of radon atoms onto soil grain surfaces [69–71].

Iskandar and Yamazawa [71] proposed an experimental mathematical correlation assessing the emanation coefficient from a soil sample at different temperatures (from −20 to 45°C):

$$f = f_0 + 0.21(T - T_0) \tag{11}$$

Where:

f_0 is the emanation coefficient measured at T_0 (°C).

The pore and grain size and the radium-226 distribution are interdependent microstructure properties that strongly influence the radon emanation.

The porosity fixed, the pore size depends on the grain size: smaller grains determine smaller pores and consequently a lower probability for radon atoms to finish the recoil in pores, that is, the radon emanation coefficient decreases [72]. The grain size also influences the specific surface area that, in turn, affects the radium distribution on grains: indeed, smaller grains (particularly of platy shape-ones, e.g., clay) have larger specific surface areas, and consequently, a larger fraction of radium atoms are concentrated near the grain surface [66, 73, 74].

The radon atoms produced close to the grain surface are more likely to enter the pore spaces than those generated inside the grain. Phong Thu and Van Thang [61] observed an increase in radon emanation coefficient at fixed moisture content for the same soil when its maximum grain size passed from >0.5 nm to <0.1 nm. If radium atoms are uniformly distributed inside the grain, the emanation coefficient decreases with increasing grain size [75]. Some authors have found an inverse proportion when the diameter exceeds 0.1 μm [39, 76]. Oppositely, if radium atoms are mainly distributed on the grain surface, the emanation coefficient increases with increasing grain size up to a plateau. The saturation of the emanation coefficient with increasing grain size was reported to occur at 100–300 μm for wet soil and 1500 μm for dry one [61, 73, 75]. Different correlations are also documented in the literature (**Table 5**) [41].

Soil description	Mean grain diameter (μm)	Porosity
Sand	60–2000	0.34–0.38
Sand and sandy clay	—	0.46–0.57
Silt	2–60	—
Clay	<2	—
Dense glacial till	1300 3000	—
Esker sand	800	—
Gravel sandy loam	—	0.35

Table 5.
Typical values reported by Font [54] of mean grain diameter and porosity.

Other influencing parameters on the radon emanation coefficient have been less investigated in the literature: the emanation properties were observed to vary according to the grain shape, that is, aggregates of small grains compared to single grains with the same size [77]. A recent work has also assessed the dependence of radon emanation on the content of Fe/Mn compounds and on the soil pH. High Fe/Mn content implies high levels of Fe/Mn oxides and oxyhydroxides so leading to a more effective radium adsorption capacity on grain surface: this determines higher radon emanation. The radium absorption capacity of the Fe/Mn oxides and oxyhydroxides is also enhanced at higher pH and so the resulting emanation coefficient [61].

2.2 Building materials

Building materials have been recognized to be a source of indoor radon since the '80s. First findings were collected by Hultqvist [78] who reported high radium-226 activity concentrations in some building materials used in Sweden, especially a type of concrete based on alum-shale. Kolb and Schmier [79] and Sorantin and Steger [80] in Austria; Krisiuk and Tarasov [81] in Eastern Europe; and Sciocchetti and Clemente [82] in Italy carried out studies with similar purposes.

In the past, building materials were considered the main source of indoor radon activity concentration [83]; only some years later, the soil was recognized to be typically the main source and the contribution of building materials to vary widely (over two orders of magnitude in absolute value) [26].

The building materials relevantly contributing to radon activity concentration indoors have both high radium-226 activity concentration and porosity degree [28]. Generally, their contribution is lower than that of soil. However, some case-studies documented building materials predominantly contributing to radon indoors (e.g., [84–86]). Some authors also reported the building material contribution to increase with the floor level (especially in large-scale buildings [87–90]), with increasing air-tightness [91], and in buildings with very thick walls [92].

2.2.1 Radium-226 activity concentration

As previously discussed about soil, the building material contribution to the indoor radon activity concentration strongly depends on the materials' radium-226 activity concentration.

Referring to **Table 6**, the last four items (in italic) are naturally occurring radioactive materials. Except for pozzolana, these are residues from different industries that since the late '80s have been used in order to move toward an economy based on reuse and recycling [94]. A wide description of the recycle of industrial wastes in building materials was carried out by Labrincha and Puertas [95].

Building material description	**^{226}Ra activity concentration**	
	AM (Bq kg^{-1})	**Range (Bq kg^{-1})**
Brick	51	7–84
Concrete	59	14–272
Cement	50	25–87
Aggregates	23	4–69

Building material description	^{226}Ra activity concentration	
	AM (Bq kg^{-1})	Range (Bq kg^{-1})
Clay	51	18–94
Chalk (i.e., $CaCo_3$)	15	9–23
Gypsum (i.e. $CaSO_4{\cdot}2\,H_2O$)	18	4–60
Lime (i.e., CaO)	19	11–30
Limestone	15	5–31
Travertine	7	—
Marlstone	35	—
Bulk stone	39	11–90
Natural stones used as superficial material	63	4–202
Fly ashes	91	75–815
Bottom ashes	345	68–1391
Tuff (Italy)	157	12–316
Tuff (Germany)	74	47–100
Granite	79	—
Basalt	81	—
Gneiss	123	—
Bauxite residues/red mud	205	97–301
By-product gypsum	318	22–668
Metallurgical slag	139	15–336
Pozzolana	187	—

Data are taken from database of activity concentration measurements of natural radionuclides in European building materials [93]. It collects data of about 23,000 samples of bulk materials from 26 of 28 European member states and 4 non-EU countries (Turkey, Macedonia, Switzerland, and Norway).

Table 6.
Arithmetic mean (AM) and range of ^{226}Ra *activity concentration* (Bq kg^{-1}) *in some building materials.*

2.2.2 Radon emanation and influencing parameters

As previously seen, the emanation coefficient represents the fraction of radon atoms produced by the radioactive decay of radium-226 that leaves the solid matrix to enter the pore volume either water- or air-filled.

Table 7 collects emanation fractions for the most used building materials including industrial by-products. Data are split in two to easily compare the findings of the past (measurements performed until the late eighties) to the most recent ones (measurements until 2018). As regards bricks, Nazaroff and Nero [29] report emanation coefficients ranging from 2 to 14%, whereas more recent findings report a wider range with an arithmetic mean of 12% [93]. The same applies to concrete. By-products like coal ash and gypsum were not included in the collection of data by Nazaroff and Nero [29]. No differences in radon emanation coefficient are found between raw gypsum and phosphogypsum, that is, the by-product resulting from the production of phosphoric acid from the phosphoric mineral.

Emanation coefficient	Range	AM	Range
	Nazaroff and Nero [29]	Trevisi, Leonardi [93]	
Concrete	0.1–0.44	0.24	0.01–0.854
Brick	0.02–0.14	0.12	0.002–0.674
Gypsum	0.03–0.24	0.13	0.03–0.244
Cement	0.02–0.05	0.12	0.007–0.564
Fly Ash	0.002–0.02	0.01	0.005–0.024
Coal Ash	—	0.005	0.004–0.009
Gypsum	—	0.13	0.03–0.214

In italic the NOR materials. Even if the values from Trevisi, Leonardi [93] have been recently updated by extending the dataset to 13 EU and 4 non-EU countries [96], such updates are not considered because the results are grouped for categories of building materials.

Table 7.
Emanation coefficient of ^{222}Rn *in building materials: Comparison between values from [29] and those from the database built by Trevisi, Leonardi [93].*

The same parameters reported to influence the radon emanation in soil (§3.1) similarly affect the emanation of radon atoms occurring in building materials.

As already introduced about the soil, the radon emanation coefficient strongly depends on radium distribution and grain size and shape. In building materials, the radium-226 is uniformly distributed in solid grains of primary minerals [97]. The distribution in secondary materials (e.g., sedimentary deposits, uranium tailings, and residues) is reported to be uniform on the grain surface [73]. For those materials undergoing high-temperature thermal production or treatment processes (e.g., coal fly ashes), the radium-226 uniform distribution on the grains surface results from the volatilization of radium occurring at high temperature and the consequent deposition on the grains surface [72].

The impact of the moisture content on the radon emanation coefficient was specifically addressed for building materials (e.g., [98, 99]). This coefficient was reported to increases as a function of moisture content up to a certain value around 10% (weight percentage) of water content.

The porosity, which significantly influences the radon emanation coefficient of building materials, including by-products [100, 101], is influenced by some characteristics of the material generation process. In case of concrete, the porosity depends on the water-cement ratio, the curing age, the aggregate type, and proportion in mixture. For materials of volcanic origin, the porosity is influenced by the time scale of the cooling process. An increase of radon emanation coefficient with porosity has been obtained also by Pereira, Lamas [102] on granite rocks and Misdaq and Amghar [103] on marble. The influence of porosity and pores specific area[4] has been studied by Jobbagy, Somlai [104] in red mud samples. A positive linear relationship was observed in the pore volume range of 0.01–0.07 $cm^3\ g^{-1}$. For higher values (i.e., >0.23 $cm^3\ g^{-1}$), Zhang et al. [64] reported a radon emanation coefficient sudden decrease.

[4] Pore specific surface of a porous material is defined as the interstitial surface area of the voids and pores either per unit mass or per unit bulk volume of the porous material.

Furthermore, some works showed the emanation coefficient does not depend only on the porosity, but on the dimensions of pores as well (e.g., [101, 105]). The mesopores (i.e., smaller than 100 nm) were recognized to be associated with high radon emanation coefficient [106]. The porosity fixed, if the pores size distribution move toward higher dimensions, a reduction of radon emanation coefficient occurs [107]. However, greater pores, although negatively influencing the emanation process, enhance the further radon transport mechanisms [106].

The general approach resulting from the literature is to avoid using the porosity as proxy of the radon emanation coefficient when comparing different types of building materials [105]. Only when discussing the difference in the radon emanation coefficient of materials of the same typology, porosity and pore specific surface are reliable proxies of the emanation coefficient [100].

A summary of typical porosity values for the most common materials used in constructions is reported in **Tables 8** and **9**.

In the past, the influence of additives in cement as enhancer of radon emanation has been also studied: no significant correlations were found [118].

Building materials	Porosity, ϵ	References
Concrete[a]	Up to 0.3	[108]
Brick[c]	0.3–0.4	[109]
Limestone[b]	Up to 0.3	[109]
Marble	0.002–0.01	[103]
Sandstone[b]	0.1–0.2	[109]
Tuff	Up to 0.4[d]	[110]
Pumice	0.7	[111]

[a]*Details on aggregates, binder, and relative proportions in the reference.*
[b]*Bulk density ranging from 1800 to 2700* kg m^{-3}.
[c]*Data are for fired clay brick.*
[d]*Raviv, Wallach [111] reports 0.6 for yellow tuff.*

Table 8.
Typical porosity of some building materials: Summary of recent literature findings.

Building materials	Pore diameter, μm	References
Basalt	0.03[a]	[112]
Clay brick	0.01–80	[113]
Clay	Up to 2	[114]
Concrete[b]	0.002–0.3	[113]
Concrete[c]	0.008–5	[115]
Concrete[d]	0.008–0.4	[115]
Granite	0,48[e]	[112]
Mudstone	0.02[f]	[112]
Limestone[g]	Up to 10[h]	[116]
Coarse-grained Sandstone	0.4–8	[117]
Fine-grained Sandstone	0.002–3.2	[117]

Building materials	Pore diameter, μm	References
Sand	60–2000	[114]
Silt	2–60	[114]
Siltstone	0.04[i]	[112]

[a]*Average pore diameter. The dominant pores concentrate between 5 and 50* nm*, whose volume accounts for about 50% of the total.*
[b]*Water to cement ration w/c = 0.7.*
[c]*Water to cement ration w/c = 0.4, age 320 days.*
[d]*Water to cement ration w/c = 0.6, age 318 days.*
[e]*Average pore diameter. The dominant pore size concentrates in the range of 303–2119* nm*, whose volume accounts for more than 40% of the total.*
[f]*Average pore diameter. The dominant pore size concentrates in the range of 9.1–90.7* nm*, whose volume accounts for more than 50% of the total.*
[g]*Five different limestones are considered by the authors.*
[h]*> 95% of open pores have a diameter lower than 10* μm.
[i]*Average pore diameter. The pore distribution is concentrated in the range of 22.7–90.7* nm*, whose volume accounts for over 70% of the total.*
The full pore size distribution is generally reported in the corresponding reference.

Table 9.
Average or range of pores size distribution of some building materials and soils.

2.3 Drinking water

Radionuclides of natural origin, including radon, are normally present in drinking water, and their activity concentration strongly varies depending on the water origin. The processes leading to the presence of radon in water are:

- the emanation of radon into water-filled pores of porous materials (i.e., rocks) from radium-226 decays occurring inside solid grains (see §2.4);
- the radioactive decay or radium-226 atoms dissolved in water [119, 120].

Exposure to radon dissolved in drinking waters derives directly from ingestion of water[5] and indirectly from the inhalation of air [5, 121, 122]. In the second scenario, the radon degasses from water depending on its solubility, and it is accumulated indoors.

Due to the relatively low solubility of radon in water, about 90% of the dose attributable to radon in drinking waters comes from inhalation rather than from ingestion [11]. Therefore, measuring the radon activity concentration in air rather than in drinking-water is suggested by WHO recommendations [122].

The International Commission on Radiological Protection has recently released the ingestion dose coefficient for radon for the first time [123]. Assuming the standard consumption rate of 2 liters per day, a committed dose of 0.1 mSv results from a radon activity concentration in water of 200 Bq L^{-1} [124].

The effective dose resulting from water ingestion is generally low compared to the exposure to airborne radon. This happens because radon activity concentration in

[5] Radon contained in drinking water can be ingested giving so dose to the lining of the stomach [121]. A definitive correlation between consumption of drinking-water containing radon and an increase of risk of stomach cancer [6] has not been stated yet. Studies carried out until now do not indicate a major effect of ingested natural radon on stomach cancer risk increase [120].

water is significantly lowered by the degassing and the natural decay occurring during the water resting, treatments, and distribution.

Two freshwater typologies are identified: the ground water – existing in underground (from near-surface to deeper than 9000 meters) water-bearing permeable rocks – and the surface water that subsists above the ground in different water bodies (e.g., mainly streams, rivers, and lakes).

Considering the overall freshwater amount available on the planet, ground ice and permafrost (not available for being directly used for human consumption, irrigation, and industrial processes) contribute for the 69.0%, lakes for the 20.9%, and rivers for the 0.5%. Swamps, water in the atmosphere, soil moisture mainly contribute for the remaining part.

Water is mainly supplied to consumers via private wells and water distribution networks. In case of water distribution system, a significant percentage of water coming from the source undergoes a purification treatment.

Groundwater plays an important role in drinking water supply (**Table 10**): about 50% of the world's human consumption is sustained by groundwater that is the primary source for 1.5–2.8 billion people [125]. In Europe, groundwater has always predominantly contributed – about 75% – to drinking water supplies [126]: the proportion of groundwater in drinking water supplies in some European countries was reported by Phok [127]. In America, about 40% of public water is supplied from groundwater. In Asia and in the Pacific area, most of households use groundwater as their primary source of drinking water in both urban and rural areas [128]. In the ten most populous African countries, the percentage of the population that rely on groundwater as their main source of drinking water ranges from 1% for Egypt to 72% for Uganda [129].

2.3.1 Radon activity concentration in drinking water

Radon activity concentration in water is highly variable. The radon levels at the consumption point depend on the water origin, that is, underground or surfaces water, and on the different processes undergoing from the uptake to the consumer, for example, storage, transportation, and treatment [130].

Groundwater is in contact with larger radon-emanating surfaces and is generally isolated from the outdoor air where radon can easily escape to. Thus, higher radon activity concentrations are much more likely to occur in groundwater than in surface water. The groundwater can be further classified according to the uptake process in spring water and water drawn via private wells and boreholes. Spring and borehole

Country	
Asia-Pacific	32%
Europe	75%
Central and South America	29%
USA	51%
Australia	15%
Africa	n.a.

Table 10.
Estimated percentage of drinking water supply obtained from groundwater, table is taken from [125].

waters are distributed in bottles or fed into the water supply network, whereas the well water is mainly drawn for private use. The short time elapsing from the extraction to the consumption is the main reason why the radon activity concentration of well water is generally higher than in both surface water and groundwater.

For both groundwater and surface water, the path covered from the source to the consumer strongly affects the resulting radon activity concentration in water at the consumption point. Longer distances are generally associated to longer times required to cover the path and higher probability for radon atoms to escape from water. Spending more time to reach the final consumer implies more decays to occur and so a lower radon activity concentration in water.

Furthermore, water treatment and distribution plant are generally responsible for a significant reduction in the radon activity concentration in water because the radon degassing is enhanced by the turbulences created in the water flux by some mechanical components, for example, pumps and filters.

The United Nations reported in 1993 typical and maximum values of radon activity concentration in well, spring, and surface water (**Table 11**), but several measurements have been more recently performed that better characterize the variability of the radon activity concentration according to the water typology and the geology of the area (**Table 12**).

Type of supply	Radon activity concentration ($Bq\ L^{-1}$)	
	Typical	Up to
Well water	100	80,000
Spring water	10	4000
Surface water	1	10

Table 11.
Typical values and upper boundaries for radon activity concentration in the well, spring, and surface water [11].

Water type	Radon activity concentration ($Bq\ L^{-1}$)	Country	Geology	Reference
Drinking water	<3	Serbia (Novi Sad)		[131]
	0.3–24	Hungary		[132]
	0.3–24	Greece and Cyprus		[133]
	1.46–644	Austria	Granite bedrock	[134]
	100		Granite bedrock	[135]
	<1.3–1800	Germany		[136]
	1.9–112.8	Portugal		[137]
	0.2–71.1	UK		[138]
	1.41–30.67	India		[139]
	1.6–46.3	North Kosovo		[140]
Surface water	<1	Slovenia		[141]
	0.5–10	Transylvania, Romania		[142]

Water type	Radon activity concentration ($Bq\ L^{-1}$)	Country	Geology	Reference
Groundwater	1–1000	Spain (La Garrotxa)		[141]
	0.2–26		Volcanic	[119]
	3043	Poland (Sudety Region)	Volcanic	[143]
	3800	Finland		[144]
	1220	Germany (East Baviera)	Granite	[145]
	17–3856	Portugal (Nisa)	Granite	[146]
	5.8–36.6	Northern Ireland	Sandstone	[147]
	3.19–171.35	India		[148]
Spring water	2–129.3	Transylvania, Romania		[142]
	2–256	Italy	Volcanic or sedimentary	[149]
Well water	10–300	Norway	Granite & Slate	[141]
	1–40	Hungary (Southern plane)		[132]
	4–63,560	Sweden (Stockholm County)		[150]
	47–1600	Belgium (Visè)		[151]
	77,000	Finland	Granite bedrock	[144]
	385–3702	Bosnia-Herzegovina (Tuzla)	Limestone	[152]
	1.5–10	Italy (Pesaro-Urbino)	Sedimentary	[153]
	0.6–112.6	Transylvania, Romania		[142]

Table 12.
Recent literature findings about radon in water activity concentration in different water types.

2.3.2 Water consumption

In 2017, The International Commission on Radiological Protection [123] indicated for the first time the committed effective dose due to radon ingestion, $6.9 \cdot 10^{-7}$ mSv Bq^{-1}.

The general approach has been to consider a standard daily consumption of 2 L of water. Recent observations have shown why this approach may lead to overestimate the public exposure due to water ingestion [124]. This happens because a fraction largely variable over the world of the daily water intake comes from bottled water (**Table 13**) that is typically characterized by lower radon activity concentrations. The activity concentration is lowered mainly by the decays occurring during the generally long time needed by the bottles to reach consumers' houses [149]. The higher radon activity concentrations are found in the tap water whose daily consumption rate is much lower than 2 L (**Table 14**). This rate is observed to generally decrease for developed countries.

Country	Consumption of bottled water in EU countries (L per person per year)
Italy	200
Germany	168
Portugal	140
Hungary	139
Spain	135
France	133
Greece	133
Belgium	130
Poland	114
Romania	106
Bulgaria	100
Austria	95
Croatia	84
Slovak Rep	73
Latvia	72
Slovenia	67
Lithuania	61
Ireland	60
Czech Rep	57
Estonia	37
UK	37
Netherlands	28
Denmark	20
Finland	17
Sweden	10
EU average	118

The top five-biggest consumers of bottled water are Italy, Germany, Portugal, Hungary, and Spain. Own elaboration from NMWE [154].

Table 13.
Consumption of bottled water per capita in the European Union in 2019.

Country	Daily consumption of tap water (L)
Canada	1.2 ± 0.8
Finland	0.5
France	0.8 ± 0.6
Netherlands	0.18
Norway	0.58

Country	Daily consumption of tap water (L)
Sweden	0.86 ± 0.48
USA	0.92
Mean	0.7

Table 14.
Daily consumption of cold tap water (L) from Chen [124].

2.3.3 Radon activity concentration in water as a source of airborne indoor radon

The following differential equation can be used to model the indoor radon activity concentration, C_i, resulting from the degassing of radon atoms contained in water.

$$\frac{dC_i}{dt} = S + C_0\lambda_v^* - C_i\lambda_v^* + \frac{C_w W^* e^*}{V} \tag{12}$$

Where:
C_0 is the outdoor radon activity concentration (Bq m^{-3}),
S is the entry rate per unit volume for all other sources (Bq m^{-3} s^{-1}),
λ_v^* is the instantaneous air exchange rate (s^{-1}),
C_w is the activity concentration of radon in water (Bq m^{-3}),
W^* is the instantaneous water-use rate (m^3 s^{-1}),
V is the volume of the cell, i.e., dwelling (m^3),
e^* is the instantaneous transfer efficiency from water to air, weighted for the different usages.

If Eq. 12 is integrated over a period long enough to suppress the periodic and aperiodic components leading to variation of indoor radon activity concentration, it results in:

$$\frac{1}{T}\int_0^T dC_i = 0 \tag{13}$$

$$\frac{1}{T}\int_0^T C_i\lambda_v^*\,dt = \frac{1}{T}\int_0^T (C_0\lambda_v^* + S)dt + \frac{1}{T}\int_0^T \frac{C_w W^* e^*}{V}dt \tag{14}$$

C is the indoor activity concentration without the radon in water contribution, and C_a^* is the radon in water contribution; Eq. 14 can be split in two, accordingly.

$$\frac{1}{T}\int_0^T C\lambda_v^*\,dt = \frac{1}{T}\int_0^T (C_0\lambda_v^* + S)dt \tag{15}$$

$$\frac{1}{T}\int_0^T C_a^*\lambda_v^*\,dt = \frac{1}{T}\int_0^T \frac{C_w W^* e^*}{V}dt \tag{16}$$

The mean air exchange rate, λ_v, is computed as follows by weighting the instantaneous exchange rate λ_v^* with the contribution to indoor radon coming from the water (C_a^*).

$$\lambda_v = \frac{\int_0^T C_a^*\lambda_v^*\,dt}{\int_0^T C_a^*\,dt} \tag{17}$$

Including Eq. 17, Eq. 16 results in:

$$\frac{1}{T}\int_0^T C_a^* \, dt = \frac{1}{\lambda_v T}\int_0^T \frac{C_w W^* e^*}{V} dt \tag{18}$$

The following time-average values are further defined:

$$C_a = \frac{1}{T}\int_0^T C_a^* \, dt \tag{19}$$

$$W = \frac{1}{T}\int_0^T W^* \, dt \tag{20}$$

$$e = \frac{\int_0^T W^* e^* \, dt}{\int_0^T W^*} \tag{21}$$

$$V = \text{const} \tag{22}$$

$$C_w = \text{const} \tag{23}$$

In Eq. 21, the transfer efficiency e^*, computed considering the different usages, is further weighted relative to the use rate (see §2.3.4). The instantaneous transfer efficiency accounts for the different transfer efficiencies of the water-involving activities (e.g., shower, dishwashing) at a certain moment. If a mean transfer efficiency is computed, the different water use rate over the time should be considered as weighting factors in place of time.

The following final formulation is obtained for radon in water contribution to the indoor radon activity concentration by including Eqs. 19–22 in Eq. 18.

$$C_a = \frac{C_w W e}{\lambda_v V} \tag{24}$$

All the previous boundary conditions fixed; Eq. 24 can be used to estimate the increase in radon indoor activity concentration due to water over a long-term period.

2.3.4 Water use rates

The distribution for the in-house activities of the per capita domestic water-use rates is reported in **Table 15** from several countries. A similar collection was provided by Nazaroff and Nero [29] who reviewed only American studies up to 1988. Data resulted to be very well fitted by a lognormal distribution with a geometric mean of 189 L person^{-1} d^{-1} and a geometric standard deviation of 1.57 L person^{-1} d^{-1}.

Water use rates are distinguished into seven categories: dishwashing, shower, bath, toilet, laundry, faucet, and all the others grouped together. This classification is from the "Residential End Uses of Water" report [156]. Bendito, Mudgal [157] referred to the same classification criteria. All the other references in **Table 15** used different classifications: data have been adapted to the above classification by applying

Ref.	Country	Dishwashing	Shower	Bath	Toilet	Laundry	Faucet	Other	Total
[155]	USA	3.8	43.9	4.5	70.0	56.8	41.2	6.0	226
[156]	USA	2.6	42.0	5.7	53.7	36.3	42.0	9.5	192
[157]	EU MS	11.4[a]	42.5	11.2	39.9	14.7	13.2	4.3	137
[158]	Denmark	14	49.0[b]	–[b]	37.0	18.0	10.0	19.0	136
[159]	Sweden	38	57.0[b]	–[b]	38.0	28.0	9.0	19.0	189
[160]	England & Wales	0.0	32.0[b]	–[b]	53.0	22.0	5.0	48.0	160
[161]	Norway	20.0	40.0[b]	–[b]	30.0	25.0	8.0	7.0	130
[162]	Finland	18.0	57.0[b]	–[b]	17.0	16.0	4.0	3.0	116
[163]	Netherlands	5.7	53.6[b]	–[b]	36.2	23.1	1.7	7.7	128
[164]	Germany	7.6	37.8[d]	–[d]	40.3	17.6	15.1[e]	7.6[c]	126
[165]	France	14.8	57.5[d]	–[d]	29.6	17.8	10.4	17.8[c]	148
[166]	Italy[f]	15.4	62.7[d]	–[d]	68.2	24.2	23.1	24.2[c]	220

[a]*The value, reported by the reference for a household, is here kept unchanged for a single person.*
[b]*The Member States communications report values for personal hygiene here attributed to shower.*
[c]*It comprehends the outdoor usages, i.e., gardening and car washing.*
[d]*Shower and bath are not distinguished in the report, and the sum is here attributed to shower.*
[e]*Obtained as the sum of personal hygiene, cleaning, and cooking/drinking.*
[f]*Use rates percentages are taken from ARPAE Emilia Romagna [167].*
Water leakages are, where explicitly distinguished, excluded from the use rate.

Table 15.
Per capita in-house water-use rates (L person^{-1} day^{-1}).

conservative criteria, that is, attributing water use rates different from those in the table to the similar usage with the highest transfer efficiency (see §2.3.5).

2.3.5 Transfer efficiency

Solubility of radon in water increases with decreasing temperature. Boyle firstly expressed the radon solubility coefficient as a function of the temperature: he found values of 0.51, 0.25 and 0.16 at 0.0, 20.0 and 39.1°C, respectively [168]. Nowadays, the radon solubility in water is expressed by the Ostwald coefficient, K. It is computed as the ratio, the thermodynamic equilibrium established, of the radon activity concentration in water, $C_{\mathrm{Rn},w}$, to the radon activity concentration in air, $C_{\mathrm{Rn},a}$, at thermodynamic equilibrium. This coefficient can be computed using the following [169, 170]:

$$\frac{C_{\mathrm{Rn},w}}{C_{\mathrm{Rn},a}} = K = 0.105 + 0.405e^{-0.0502\,T} \tag{25}$$

where T is expressed in Celsius degrees (°C) (**Figure 7**).

The time required to spontaneously reach thermodynamic equilibrium is shortened by some phenomena provoking artificial degassing, for example, water agitation and bubbling.

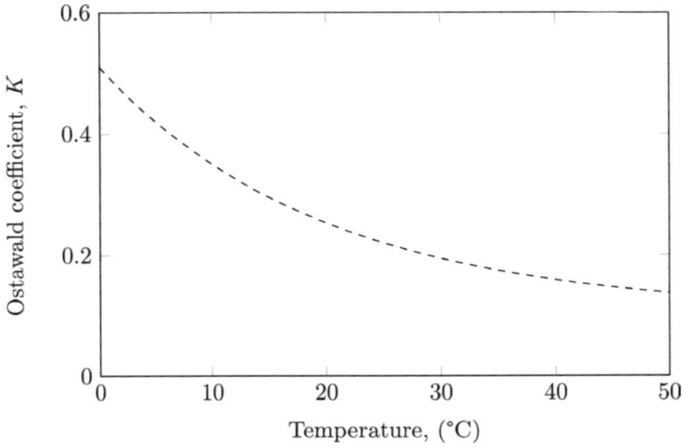

Figure 7.
Ratio of the radon activity concentration in water to the radon activity concentration in air at thermodynamic equilibrium as a function of water temperature.

Type of use	Partridge, Horton [171]	Gesell and Prichard [172]	Hess, Weiffenbach [173]
Dishwashing	0.98[a]	0.90	0.98
Shower	0.71	0.63	0.65
Bath	0.60[b]	0.47	0.30[e]
Toilet	0.29[c]	0.30	0.30[e]
Laundry	0.98[d]	0.90	0.90[e]
Faucet	0.28	0.45	0.30[e]

[a]*Arithmetic mean of values for shower with cold and hot water.*
[b]*Value obtained for bathing with hot water.*
[c]*0.24 for bowl and 0.5 for tank.*
[d]*Value obtained for hot (or warm) wash cycle lasting 18 minutes.*
[e]*Values estimated in the reference.*

Table 16.
Transfer efficiency for the release of radon from water to air, by domestic use.

However, some experimental evidence showed that the thermodynamic equilibrium is generally not attained in actual scenarios: so, the actual radon fraction released indoor is generally much smaller than that computed according to Eq. 25. The difference between the situation at the thermodynamic equilibrium and that in the actual scenario is accounted by the transfer efficiency, e. It stands for the ratio of the radon activity (or activity concentration) in water after a given time depending on the specific usage to the radon activity (or activity concentration) in the same water before the usage (**Table 16**).

The common approach to assess the exposure to indoor radon released from the water is to consider the entire house and to integrate over periods of some days (or even more) [174]. Under such conditions, the transfer of radon from the water into the indoor air is commonly modeled by a transfer coefficient C_T.

$$C_T = \frac{\Delta C_{Rn,a}}{C_{Rn,w}} \tag{26}$$

Several surveys showed C_T to be distributed over a wide range approximately centered in 10^{-4} [173–176].

2.4 Conclusion

The increased – and more consolidated than in the past – knowledge on the risks associated to radon exposure has led the attention to be focused on the development of radon reduction strategies in buildings more and more specific and effective. This has reflected on the widespread of both practical and theoretical approaches to assess the contributions of the different sources to the indoor radon activity concentration. Most of the theoretical approaches consist of modeling the radon entries mainly from the soil and, in few cases, from the building materials.

The resulting large availability of mathematical formulations of the radon generation processes has made clear the lack of a harmonization and standardization of both the parameters regulating, or influencing, the radon availability, and the assumptions underlying the formulations typically considered by the authors. Several examples exist of parameters either defined differently or referred to by various symbols and of mathematical formulations being used improperly without considering the actual scenarios.

Furthermore, all the models provided to assess the indoor radon concentration requires a large set of inputs to be known. Most of these are generally difficult to measure, so the common approach consists of considering typical values available in literature. In the last years, many measurements have been carried out on the different parameters affecting the radon generation process, but no systematic review exists.

This chapter considers the contributions of several authors that have been working in the recent years on the modeling of radon generation mechanisms. The results are the homogenization of the current theoretical knowledge about radon generation and the standardization of both parameters and assumptions influencing the modeling. Finally, large reviews have been carried out to report the typical values of most of the parameters regulating or influencing the radon generation.

3. Radon transport

In porous media with negligible moisture content, the radon activity concentration can be modeled through the following general transport equation, with vectors in bold [29]:

$$\frac{dC_{\mathrm{Rn}}}{dt} = -\nabla \cdot \boldsymbol{J}^{d}_{\mathbf{Rn}} - \nabla \cdot \boldsymbol{J}^{a}_{\mathbf{Rn}} - \lambda_{\mathrm{Rn}} C_{\mathrm{Rn}} + G_v \tag{27}$$

The first and second terms on the right-side of Eq. 27 account for the variation of radon activity concentration by diffusive and convective transport, respectively. The variations of radon activity concentration due to diffusive and convective transport can be expressed through the Fick's (see §3.1) and Darcy's law (see §3.2), respectively. It results in:

$$\frac{dC_{\mathrm{Rn}}}{dt} = D_e \nabla^2 C_{\mathrm{Rn}} + \frac{k}{\mu \varepsilon} \nabla \boldsymbol{P} \cdot \nabla \boldsymbol{C}_{\mathbf{Rn}} - \lambda_{\mathrm{Rn}} C_{\mathrm{Rn}} + G_v \tag{28}$$

Where the vectors are in bold and:

D_e is the effective diffusivity of porous medium along the x-axis ($m^2\ s^{-1}$),

ε is the medium porosity,

C_{Rn} is the activity concentration of radon per unit volume of interstitial space ($Bq\ m^{-3}$) and $\nabla \mathbf{C_{Rn}}$ the corresponding gradient vector,

λ_{Rn} is the radon decay constant (s^{-1}),

$\nabla \mathbf{P}$ is the absolute pressure gradient vector ($Pa\ m^{-1}$),

k is the permeability of the porous material (m^2),

μ is the carrying fluid (i.e., mainly the air) dynamic viscosity (Pa s),

G_v is the radon production rate, per unit pore volume ($Bq\ s^{-1}\ m^{-3}$).

Although thermodiffusion was reported to be a likely transport mechanism in some specific circumstances [177], Eq. 27 refers to the radon transport mechanisms presented and discussed in the latest UNSCEAR publications [26, 28].

Eq. 28 relies on some simplifying assumptions: (i) water content is assumed negligible, that is, all pores are air filled, so the migration of radon occurs in air; (ii) no adsorption of radon atoms is assumed to occur on the surfaces of the solid grains; and (iii) radon is assumed to migrate only down its activity concentration gradient and/or the air pressure gradient. Furthermore, porous media are supposed to be isotropic and homogeneous relative to diffusion coefficient, permeability, porosity, emanation coefficient, radium content, and bulk density.

Removing the isotropy and homogeneity of the porous medium, and the negligibility water content, Eq. 28 turns into the following formulation to be used to model the radon transport with the only assumption to neglect the radon transport in water.

$$\frac{1}{\varepsilon}\frac{d(C_{Rn,a}\varepsilon_a + C_{Rn,w}\varepsilon_w)}{dt} = \nabla\cdot\nabla(\mathbf{D}C_{Rn,a}) + \nabla\cdot\left(C_{Rn,a}\frac{\mathbf{k}}{\mu}\cdot\nabla\mathbf{P}\right) - \frac{1}{\varepsilon}\lambda_{Rn}(C_{Rn,a}\varepsilon_a + C_{Rn,w}\varepsilon_w) + G_v \tag{29}$$

Where the vectors are in bold and:

C_a and C_w are the radon activity concentrations in the gas- and water-filled pores, respectively.

ε is the medium porosity composed by ε_a and ε_w, which are the gas and water porosity, respectively, that is, the ratio of the gas- or water-filled pores to the total pore volume.

Furthermore, D_e, k and G_v need to be estimated considering the influence of water content (see sections §2.1.2, §2.2.2, §3.1.1, and §3.2.1).

The radon production rate, G_v, inside the material can be neglected when the radium-226 activity concentration or the radon emanation coefficient are such that the number of radon atoms available in pores for the further transport is negligible. The radon production rate can be computed as follows if the radon atoms in water-filled pores are assumed negligible (see §3.1).

$$G_v = \frac{1}{\varepsilon}C_{Ra}\rho f \lambda_{Rn} \tag{30}$$

Where:

C_{Ra} is the activity concentration of radium-226 per unit mass of the material ($Bq\ kg^{-1}$),

ρ is the density of the material ($kg\ m^{-3}$),

f is the emanation coefficient,

λ_{Rn} is radon decay constant (s^{-1}).

The factor $\frac{1}{\varepsilon}$ is introduced to move from the radon production rate per unit bulk volume to the corresponding rate per unit pore volume. Other authors consider $\frac{1-\varepsilon}{\varepsilon}$ in place of $\frac{1}{\varepsilon}$ [54].

The radon activity concentration function inside the porous material can be derived by solving Eq. 28 considering a set boundary conditions specific for the scenario considered. Thus, radon exhalation rate, that is, radon atoms coming out from the interface dividing the material and the outdoor air, per unit time and surface, can be obtained as:

$$E(T) = E(T)^{\text{diff}} + E(T)^{\text{adv}} = -D\frac{dC_{\text{Rn}}(x)}{dx}\bigg|_{x=T} - \frac{C_{\text{Rn}}(x)}{\varepsilon\mu}\frac{k\,dp}{dx}\bigg|_{x=T} \tag{31}$$

Where $x = T$ is the abscissa at the surface dividing the material in contact with the environment to which radon atoms are exhaled.

3.1 Diffusive transport

The diffusive transport refers to the migration of a molecular species down its activity concentration gradient [29].

Radon diffusion from the soil through the floor material, mainly concrete, can be a significant mechanism for radon entry into dwellings [28], typically after advection transport from soil (see **Table 1**). This is because the quality of concrete in floor slabs is not so high, and the porosities are significant. For the same reason, radon diffusion transport through building materials with a combination of high ^{226}Ra content, and porosity is enhanced.

The Fick's Law governs such a molecular motion through the so-called diffusion coefficient (sometimes referred to as molecular diffusivity). The diffusion coefficient in open air, D_0, needs to be adjusted when addressing migration in porous media to consider two different phenomena:

i. the presence of solid particles (in porous media generally referred to as "grains") causes the diffusion paths of species to deviate from straight lines. The diffusion coefficient must be adjusted to account for the deviations by the tortuosity factor defined as the actual distance traveled by the species (l) per unit length of the medium crossed (x), $\tau = \frac{\Delta l}{\Delta x}$ [178]. Dullien [179] defines the tortuosity factor as $\tau = \frac{\Delta l^2}{\Delta x^2}$.

ii. The presence of the grains reduces the cross sectional area through which the diffusion occurs. This reduction is accounted by a fraction equal to the ratio of the open pore area, A^*, to the total cross section, A, that is, the areal porosity (**Figure 8**) [180].

The bulk coefficient, in the following referred to as D, is introduced to account for the lengthening of the path traveled (i). It relates the gradient of interstitial activity concentration of diffusing radon (becquerel per cubic meters of pore volume) to the flux density over the cross-sectional area (becquerel per square meters of the cross-section). According to Shen and Chen [178]:

Figure 8.
Open pore area at a given cross-section of a porous material: Rearrangement from Culot, Olson [181].

$$D = D_0 \tau^2 \tag{32}$$

In case of high porous media, τ was found to be well fitted by the porosity ε, so $\tau^2 \approx \varepsilon^2$ [182]. It results in [183]:

$$D = D_0 \varepsilon^2 \tag{33}$$

Other correlations have been proposed for $\tau(\varepsilon)$ [178, 184].

The effective coefficient, in the following referred to as D_e, is introduced to account for the reductions in cross-sectional area (ii), and it relates the gradient of interstitial activity concentration of diffusing radon (becquerel per cubic meters of pore volume) to the flux density over the cross-sectional pore area (becquerel per square meters of the cross section). D_e is related to D by the following relationship:

$$D_e = D \frac{A}{A*} \approx D \frac{V}{V_v} = \frac{D}{\varepsilon} \tag{34}$$

Where:

V is the overall volume of the porous material (m^3),

V_v is the void volume of the porous material, both water- and gas-filled (m^3),

ε is the material porosity.

The assumption underlying Eq. 34 is the equality between the areal porosity (i.e., the fraction of open pore area in a unit cross-section) and the volume porosity, that is, $\frac{A}{A*} \approx \frac{V}{V_v} = \varepsilon$ [181]: the assumption is legit in case of porous media with random structure [179].

Rather than the diffusion coefficient, some authors prefer using the diffusion length, R, or its inverse, r (e.g., [185]):

$$R = \sqrt{\frac{D_e}{\lambda_{\mathrm{Rn}}}} \tag{35}$$

$$r = \frac{1}{R} = \sqrt{\frac{\lambda_{\mathrm{Rn}}}{D_e}} \tag{36}$$

The diffusion of radon in isotropic and homogeneous porous media is modeled by the first Fick's law:

$$\boldsymbol{J}^d_{\mathbf{Rn}} = -D_e \nabla \boldsymbol{C}_{\mathbf{Rn}} \tag{37}$$

Where the vectors are in bold and:

$\boldsymbol{J}^d_{\mathbf{Rn}}$ is the diffusive flux density vector of radon activity per unit pore area of the material (Bq m^{-2} s^{-1}),

D_e is the effective diffusion coefficient [178] (m^2 s^{-1}),

C_{Rn} is the activity concentration of radon per unit volume of interstitial space (Bq m^{-3}) and $\nabla \boldsymbol{C}_{\mathbf{Rn}}$ the corresponding gradient vector.

Radon migration can be modeled by Eq. 37 under two main assumptions:

i. all the kinetic interactions of radon atoms happen as in the open air, that is, with other gas molecules and not with the solid boundaries of grains [29]. The reasonability of this assumption depends on the recoil range of radon atoms [40, 186] relative to the dimension of open pores. When pore diameters are mostly lower than the recoil range of radon, as it happens for most building materials, the gaseous atoms collide with the wall rather than with other atoms. The resulting diffusion coefficient of radon in open air (D_0) is substituted by the Knudsen diffusivity – strongly dependent on the position within the pores [187] – to consider the pores diameter [188].

ii. All the radon atoms are entirely either in the air filling the voids or in the solid matrix [29]. This assumption requires the pores size distribution to be unimodal, the fraction of radon contained in water-filled pores to be negligible, and no appreciable adsorption of radon atoms on solid grains to happen.

If the medium is assumed to be not isotropic and homogeneous, the diffusion coefficient is no longer uniform over the medium, and diffusion transport may be modeled by Fokker–Planck diffusion law [189]:

$$J^d_{\mathrm{Rn}} = -\nabla(\boldsymbol{D_e} C_{\mathrm{Rn}}) \tag{38}$$

Where $\boldsymbol{D_e}$ is the diffusivity tensor.

3.1.1 Diffusivity

The molecular diffusivity, usually referred to as diffusion coefficient, is the key parameter of the diffusion process. It can be interpreted as the tendency of a substance to migrate down its activity concentration gradient in a material.

No standardization currently exists about terminology and symbols used to denote the diffusion coefficient: it is defined in more than four ways depending on the reference volume of the radon activity concentration – bulk or pore volume – and of the resulting flux density.

The bulk diffusion coefficient (in the following referred to as D but sometimes as k_e, (e.g., [181]) relates the interstitial activity concentration of the diffusing species (i.e., radon) to the flux density across the geometrical cross sectional area, A (see **Figure 8**). The effective diffusion coefficient (in the following referred to as D_e but sometimes as k_e^*, (e.g., [181]) relates the same interstitial activity concentration of D to the flux density across the pore area, A^* (see **Figure 8**).

The upper bound of the radon diffusion coefficient in porous media is the radon diffusion coefficient in open air, $D_0 = 1.2 \cdot 10^{-5}\,\mathrm{m^2\,s^{-1}}$ [190]. The diffusive radon flux density through a porous medium is lower than in a homogeneous medium (e.g., air) for two main reasons: (i) a smaller volume is available for diffusion, (ii) the path traveled by the diffusing species is tortuous (so longer) around solid grains. The tortuosity of the resulting path traveled by the diffusing species is accounted by the tortuosity factor[6], always higher than unity in porous material [192].

A quite large database collecting data for diffusion coefficient in different soils is given by Nazaroff [191]. A typical value of radon effective diffusion coefficient in low moisture content (up to 50%) soil is $10^{-6}\,\mathrm{m^2\,s^{-1}}$ (**Tables 17** and **18**) [34].

Porosity, grain size, and moisture content affect the effective diffusive coefficient in porous media by operating on the air volume available to radon for migration [61, 195, 196]. Considering each of them separately:

- as porosity decreases, the void volume inside the medium proportionally decreases, and the radon diffusion coefficient is consequently reduced,

- as grains size decreases, the pore gap length decreases, and the radon diffusion coefficient is consequently reduced,

- as saturation degree increases, the fraction of void volume not water-filled decreases (i.e., radon atoms must travel a longer path through the air-filled pores or diffuse through water), and the radon diffusion coefficient is consequently reduced.

The effect of the moisture content is significant, especially at high moisture content (see **Figure 9**) [34, 61], whereas at lower saturation degrees, water molecules form a thin and discontinuous film around the soil particles with little influence on radon diffusion.

The effect of the grain size on the diffusion coefficient decreases gradually up to about 0.1 mm [61].

Rogers and Nielson [197] proposed a well-fitting correlation that can be used to estimate the radon diffusion in "earthen materials" as a function of porosity, ε, and saturation degree, S:

$$D_e = D_0\, \varepsilon\, e^{\left(-6S\varepsilon - 6S^{14\varepsilon}\right)} \tag{39}$$

---

[6] The diffusive tortuosity, τ_d, is computed as the squared ratio of the average length of a diffusive pathway, $\overline{L}_d$, to the straight-line length, L_s [191].

Soil description	D_e ($m^2 s^{-1}$)	Comments
Mill tailings	$(6.3 \pm 0.9) \cdot 10^{-6}$	Moisture content 0.7–1.5% dry weight
Eluvial-detrital granodiorite	$4.5 \cdot 10^{-6}$	Dry
Silty sandy clay	$2.7 \cdot 10^{-6}$	Moisture content 1.5% dry weight
	$2.5 \cdot 10^{-7}$	Moisture content 10.5% dry weight
	$6.0 \cdot 10^{-8}$	Moisture content 17.3% dry weight
Compacted silty sands	$(3.0 \pm 1.3) \cdot 10^{-6}$	Porosity = 0.29–0.36. Saturation = 0.05–0.34.
Compacted clayey sands	$(3.2 \pm 1.5) \cdot 10^{-6}$	Porosity = 0.32–0.39. Saturation = 0.09–0.55.
Compacted inorganic clays	$(2.5 \pm 1.0) \cdot 10^{-6}$	Porosity = 0.32–0.43. Saturation = 0.06–0.34.
Diluvium of metamorphic rocks	$1.8 \cdot 10^{-6}$	Dry
Eluvial-detrital deposits of granite	$1.5 \cdot 10^{-6}$	Dry
Loams	$8 \cdot 10^{-7}$	Dry
Varved clays	$7 \cdot 10^{-7}$	Dry
Mud	$5.7 \cdot 10^{-10}$	37% moisture content
Mud	$2.2 \cdot 10^{-10}$	85% moisture content

Table 17.
Effective diffusion coefficient of radon in some soils, taken from [29].

Diffusion length (10^{-3} m)	Range	AM	n°	Range	AM
	Nazaroff and Nero [29]	Keller, Hoffmann [193]			Narula, Chauhan [194]
Gypsum	800–1300	1060	(12)	790–1280	1100
Heavy concrete	–[a]	60	(8)	20–7044	–[a]
Aerated concrete	–[a]	790	(5)	690–9804	–[a]
Brick	200–4004	410	(8)	220–4904	–[a]
Sandstone	–[a]	1020	(3)	930–1100	–[a]
Limestone	–[a]	400	(5)	310–4904	–[a]
Pumice	–[a]	840	(8)	650–1050	–[a]
Granite	–[a]	–[a]	–[a]	–[a]	170
Cement	–[a]	–[a]	–[a]	–[a]	760
Wall putty	–[a]	–[a]	–[a]	–[a]	69
Fly ash	–[a]	–[a]	–[a]	–[a]	980

[a]Nazaroff and Nero [29] simply refer to "concrete", $R \in [0.06 - 0.2]$ m.
Between brackets the number of samples considered by Keller, Hoffmann [193]. Narula, Chauhan [194] considered two samples for each material. In italic the by-product building materials.

Table 18.
Radon diffusion length in building materials: Comparison between values from Nazaroff and Nero [29] and those from the most recent studies on a wide set of samples for each material [193, 194].

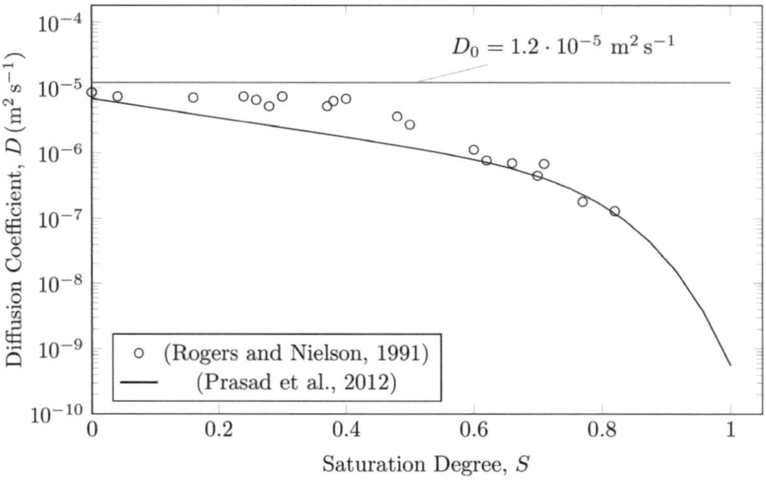

Figure 9.
Comparison between experimentally measured values by Prasad, Ishikawa [34] and those computed by the equation proposed by Rogers and Nielson [197]. Porosity assumed in the equation equals the one found for measured samples, $\varepsilon = 57\%$. Similar results were found by Phong Thu, Van Thang [61].

As soil temperature increases, the radon diffusion coefficient increases as well with a logarithmic trend. Phong Thu et al. [61] showed the radon diffusion to be 7 times stronger passing from 25 to 100°C. At low soil temperature (below the freezing point), the diffusion coefficient of wet soils decreases by a factor 2, whereas that of dry soils remains rather constant [198].

3.2 Convective transport

The convective (in literature, "advective" is commonly used to denote the same mechanism) transport describes the migration of molecular species via a bulk motion governed by fluid's head difference. The convective transport is mostly driven by a pressure gradient because the effects of gravity are generally negligible for gasses in porous media.

As seen, most of radon entering indoor comes from the ground due to the pressure difference that exists between inside and outside (see **Table 1**). This pressure difference occurs mainly due to three coexisting effects:

- the chimney effect that establishes an under pressure inside a heated building,
- the wind effect that produces local depressurization close to building boundaries due the high wind velocities on them,
- the "vacuum" effect that enhance the under pressure in the building interior due to the operation of some domestic devices, for example, mechanical ventilation systems.

Some technical regulations on buildings require indoors to be maintained under negative pressure to avoid moisture condensation on building envelope likely to result in damaging it [199, 200]. Furthermore, stack effect, wind effect, and natural ventilation usually establish an under pressure inside the building of the order of 5–10 Pa. This gradient generally increases during the heating season up to 15 Pa [201] or if energy efficiency measures are applied in the building [199].

Atmospheric pressure and temperature fluctuations, as well as other meteorological events such as snow and rainfall, may lead to an increase of soil pressure, enhancing the convective radon transport into building [202].

The governing equation of advection in isotropic and homogeneous porous media is the Darcy's law [203, 204]:

$$\boldsymbol{\nu} = -\frac{k}{\mu}\nabla \boldsymbol{P} \tag{40}$$

Where the vectors are in bold and:

$\boldsymbol{\nu}$ is the volumetric fluid flow rate vector ($cm^3\ s^{-1}$) per unit geometrical area (cm^2), also referred to as superficial velocity vector, defined over a region large relative to individual pores but small to the overall dimensions of the material ($m\ s^{-1}$),

$\nabla \boldsymbol{P}$ is the pressure gradient vector inside the porous media ($Pa\ m^{-1}$),

k is the permeability of the porous material (m^2),

μ is the carrying fluid dynamic viscosity (Pa s).

The Eq. 40 is valid only if the soil permeability is isotropic (i), the effects of gravity are negligible (ii), and the flow through the porous material happens as a viscous flow through a pipe (iii).

If the soil permeability is not isotropic, that is, k becomes a 3x3 matrix, Darcy's law may be formulated as:

$$\boldsymbol{\nu} = -\frac{\boldsymbol{k}}{\mu}\cdot \nabla \boldsymbol{P} \tag{41}$$

Given the superficial velocity vector, the advective transport of radon in porous media is modeled as follow:

$$\boldsymbol{J}^{a}_{\mathbf{Rn}} = \frac{C_{Rn}\boldsymbol{\nu}}{\varepsilon} \tag{42}$$

Where the vectors are in bold and:

$\boldsymbol{J}^{a}_{\mathbf{Rn}}$ is the advective flux density vector of radon activity per unit pore area of the material ($Bq\ m^{-2}\ s^{-1}$),

C_{Rn} is the activity concentration of radon per unit volume of interstitial space ($Bq\ m^{-3}$),

ε is the porous material porosity.

The advective term in Eq. 28 is obtained by considering the advective transport term of Eq. 42 with in the hypotesis of soil isotropy and homogeneity (see Eq. 40).

$$-\nabla\cdot \boldsymbol{J}^{a}_{\mathbf{Rn}} = \nabla\cdot\left(\frac{C_{Rn}}{\varepsilon}\boldsymbol{v}\right) = \nabla\cdot\left(\frac{C_{Rn}}{\varepsilon}\frac{k}{\mu}\nabla \boldsymbol{P}\right) = \nabla\left(\frac{C_{Rn}}{\varepsilon}\frac{k}{\mu}\right)\cdot\nabla \boldsymbol{P} + \left(\frac{C_{Rn}}{\varepsilon}\frac{k}{\mu}\right)\nabla\cdot\nabla \boldsymbol{P} \tag{43}$$

Considering the air as an incompressible fluid in the range of pressures of interest, at steady state the Laplace equation is satisfied [29], so:

$$\nabla \cdot \nabla \boldsymbol{P} = \nabla^2 \boldsymbol{P} = 0 \tag{44}$$

$$-\nabla \cdot \boldsymbol{J}^a_{\mathbf{Rn}} = \nabla \left(\frac{C_{\mathrm{Rn}}}{\varepsilon} \frac{k}{\mu} \right) \cdot \nabla \boldsymbol{P} = \frac{k}{\mu\varepsilon} \nabla \boldsymbol{C}_{\mathbf{Rn}} \cdot \nabla \boldsymbol{P} \tag{45}$$

The Darcy's law has been extensively used in applications at Reynolds (Re) number [205] up to 1 [206], and a general consensus exists about Darcy's law applicability for an upper limit of Re value between 1 and 10 for average grain size and velocity [207]. For higher Re numbers, due to emerging importance of inertial forces, deviations of fluid flow from Darcy's law have been observed even if the flow keeps laminar [208–210]. For flows in porous media with Reynolds numbers greater than about 1 to 10, the inertial Forchheimer term [209] should be added to the Darcy's equation.

The applicability of the Darcy's law requires the pores to be large relative to the mean free path of the gas to assume the particle-wall interactions as negligible [211].

3.2.1 Permeability

Soil permeability can be defined as the ease with which fluids (gases or liquids) penetrate or pass through a bulk mass of soil or a layer of soil [212]. When dealing with the radon transport, the permeability relates the advective flow of the radon-bearing air within a porous media to the pressure gradient.

Permeability is usually expressed in Darcy[7] (D) or in squared meters (m^2).

The permeability is recognized to be influenced by the soil grains size and shape, as well as by the medium porosity and the moisture content [191].

Considering each of them separately:

- as porosity decreases, the air flow through the medium becomes more difficult because less connections patterns are available, and the permeability is consequently reduced,

- as grains size decreases, the frictional resistance encountered by the air increases, and the permeability is consequently reduced,

- as saturation degree increases, less air-filled pores are available for air migration, and the permeability is consequently reduced.

The air permeability is observed to be roughly constant with increasing saturation degree up to 0.4. When exceeding this value (sometimes defined for soil as the "field capacity" [29]), large pores begin to be filled with water [29], so the air permeability decreases to 0 reached when all the pores are completely water-filled [191, 197]. The saturation degree of the soil is influenced by the rainfalls: during rainy periods, the so-called "capping effect" causes radon to be trapped in soil under a water-saturated layer characterized by low air permeability. Snow and ice on soil during winter have a

---

[7] A medium with a permeability of 1 Darcy permits a flow of 1 $cm^3\ s^{-1}$ of a fluid with viscosity 1 mPa s under a pressure gradient of 1 atm cm^{-1} acting across an area of 1 cm^2. 1 Darcy corresponds to about 1 μm^2.

similar effect. Moreover, a strong reduction of permeability is observed in frozen soil; this effect increases with increasing soil moisture content [213].

Many other parameters, hereafter not discussed in detail, influence soil air permeability: (i) the presence of fissures, cracks, and other structural void spaces [214], (ii) the anisotropy of soil permeability (e.g., quite common in sedimentary beds deposited in layers) generally with strong differences between horizontal and vertical permeability, and (iii) the presence of bio pores, that is, penetration in the soil due to animals or plants activity.

Scheidegger [215] reported a formulation to estimate the permeability moving from the material geometrical properties.

$$k = \frac{c\varepsilon^3}{TS^2} \tag{46}$$

Where:

c is a constant that depends on the pores shape, ranging between 0.5 and 0.67,

T is the medium tortuosity,

ε is the medium porosity,

S is the specific surface area defined, for spherical particles, as $S = \frac{6(1-\varepsilon)}{d}$, with d is the average grains diameter.

A widely used empirical correlation to predict the soil air permeability is reported by Rogers and Nielson [197]:

$$K = \left(\frac{\varepsilon}{500}\right)^2 d^{\frac{4}{3}} e^{-12S^4} \tag{47}$$

Where:

ε is the soil porosity,

d is the arithmetic mean of the grains diameter (m),

S is the soil saturation degree.

The Joint Research Centre (JRC) of the European Commission has recently made available online, in the framework of the European Atlas of natural radiation project [57], a European map of the soil permeability (**Table 19**) [216].

Permeability typical values and ranges are reported in **Table 20** and **Table 17**.

Permeability (m^2)	Soil typology
10^{-7}–10^{-9}	Clean gravel
10^{-9}–10^{-12}	Clean sands Clean sand and gravel mixtures
10^{-12}–10^{-16}	Very fine sands Organic and inorganic silts Mixtures of sand, silt, and clay Stratified clay deposit
$<10^{-16}$ m^2	Impermeable soils, e.g., homogeneous clays below the weathering zone

The values are taken from Nazaroff and Nero [29].

Table 19.
Permeability ranges of representative soil typologies.

Soil description	Typical permeability (m^2)	Reference
Well-graded gravel	$1 \cdot 10^{-8}$	[217]
Esker sand (dried)	$4 \cdot 10^{-11}$	[54]
Gravely sandy loam	$1.11 \cdot 10^{-11}$	[54]
Sand	$5.45 \cdot 10^{-11}$	[54]
Uniform coarse sand	$1 \cdot 10^{-9}$	[217]
Uniform medium sand	$1 \cdot 10^{-10}$	[217]
Clean, well-graded sand and gravel	$1 \cdot 10^{-11}$	[217]
Uniform fine sand	$1 \cdot 10^{-12}$	[217]
Sand and sandy clay	$3 \cdot 10^{-12}$	[54]
Well-graded silty sand and gravel	$1 \cdot 10^{-13}$	[217]
Silty sand	$1 \cdot 10^{-13}$	[217]
Uniform silt	$1 \cdot 10^{-14}$	[217]
Sandy clay	$1 \cdot 10^{-15}$	[217]
Sandy clay, loam	$1 \cdot 10^{-11}$	[54]
Dense glacial till	$1 \cdot 10^{-14}$	[54]

The values are taken from Scott [217] and Font [54] reviews on typical permeability values in soils.

Table 20.
Permeability typical values for some common soil typologies.

The permeability of the concrete mainly depends on the cement paste and the interfacial transition zone (ITZ[8]) permeability (**Table 21**). The cement paste permeability is mainly influenced by the water-cement ratio that controls the cement microstructure and the porosity of the hydrated cement paste. The water-cement ratio affects the permeability in two ways: when the water-cement ratio decreases, the cement becomes more prone to cracking, and consequently, the permeability increases [221], but at the same time, the cement porosity decreases, and so does the permeability. These two concomitants and competing effects define an optimum value of water-cement ratio that maximizes the permeability.

The ITZ locally represents a weaker and more permeable zone relative to the bulk cement past [222]. More and greater ITZs are associated to higher concrete permeability. The influence of the ITZ on the overall permeability increases with increasing aggregates volume [223]. Increasing the aggregates volume results in an increase of the concrete permeability, whereas the usage of supplementary cementing materials (such as silica fume, fly ashes, and pozzolan) leads to the reduction of aggregates fraction in paste, so the concrete permeability decreases [224].

Furthermore, the concrete permeability usually decreases with the cement age [224].

Renken and Rosenberg [225] documented permeability measurements performed on three concrete mix: (i) a typical concrete for basement slab ($\varepsilon = 0.12$), (ii) a concrete obtained by substituting in the first mix one fourth of the Portland cement

[8] This is the region of the cement paste around the aggregate particles.

Building materials	Air permeability (m^2)			
	Mean	Minimum	Maximum	References
Lime-silica Brick[a]	$3.9{\cdot}10^{-8}$	—	—	[218]
Clay Brick[a]	$6{\cdot}10^{-9}$	—	—	[218]
Sandstone[a]	$1{\cdot}10^{-9}$	—	—	[218]
Concrete	10^{-14}—10^{-16}	—	—	[219]
Polymer insulating materials	$<10^{-16}$	—	—	[219]
Insulation materials[b]	—	$1.5{\cdot}10^{-16}$	$4.3{\cdot}10^{-9}$	[220]
Wood[b]	—	$1.5{\cdot}10^{-17}$	$7.3{\cdot}10^{-16}$	[220]
Brick[b]	—	$2.2{\cdot}10^{-15}$	$2.1{\cdot}10^{-14}$	[220]
Concrete and mortar[b]	—	$2.9{\cdot}10^{-16}$	$1.9{\cdot}10^{-12}$	[220]
Gypsum board[b]	—	$4.6{\cdot}10^{-16}$	$8.7{\cdot}10^{-14}$	[220]
Cement board[b]	—	$4.4{\cdot}10^{-17}$	$4.4{\cdot}10^{-13}$	[220]
Stone (lime, sand, granite)[b]	—	$1.5{\cdot}10^{-16}$	$8.7{\cdot}10^{-15}$	[220]

[a]*Quenard et al. (1998) measured the air permeability of two types of brick and a natural sandstone used as building material. The first brick is formed by a high-pressure steam curing of a mixture of lime and silica (open porosity of 30%). The second brick is a hard-burned clay brick (open porosity of 20%). The air permeability of a sandstone (open porosity of 22%) was measured as well.*
[b]*Permeability is expressed by the authors in* $kg\ m^{-1}\ s^{-1}\ Pa^{-1}$. *The values are converted in* m^2 *multiplying by* $\frac{\mu_{air}\,(Pa\ s)}{\rho_{air}\ (kg\ m^{-3})}$.

Table 21.
Permeability of some buildings materials used by the construction industry.

with fly ash ($\varepsilon = 0.20$), and (iii) a mix with an increased water-to-cement ratio ($\varepsilon = 0.17$). Permeability values of $1.35 \cdot 10^{-16}$ m^2, $4.97 \cdot 10^{-16}$ m^2 and $3.00 \cdot 10^{-16}$ m^2 are reported for the three typologies.

Rogers and Nielson [226] performed measurements – obtaining similar results – on concrete samples of porosity ranging from 0.17 to 0.26 (**Figure 10**).

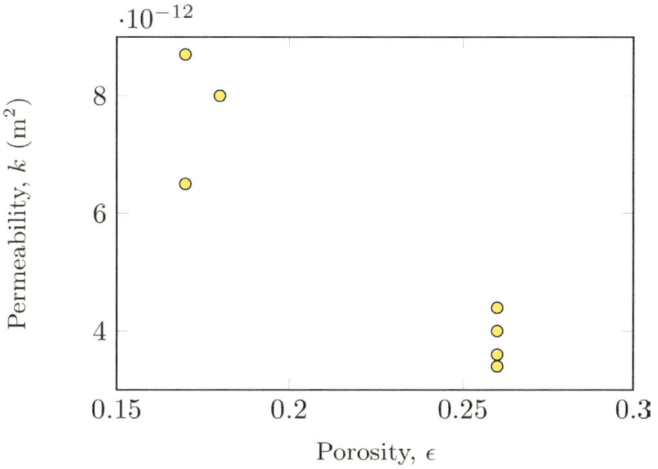

Figure 10.
Experimental results from Rogers and Nielson [226] for air permeability of concrete samples.

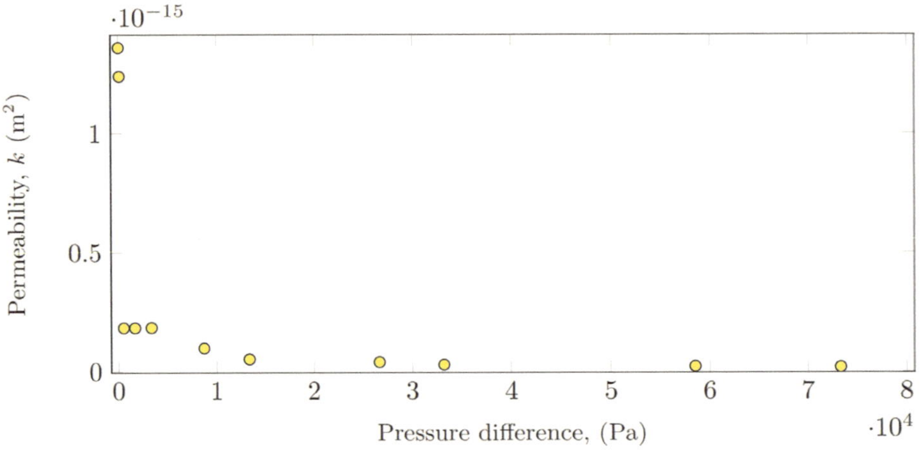

Figure 11.
Variation of radon permeability with pressure differential.

The influence of porosity on permeability of cementitious materials has been addressed also by Sakai [227]. Air permeability for concrete were reported to be in the range 10^{-13}–10^{-18}, and a poor correlation between air permeability coefficient and porosity was found.

Finally, the permeability is significantly affected by the pressure difference [202] (**Figure 11**).

Typical air permeability values for some building materials are shown in **Table 20**.

4. Conclusion

Observations similar to those previously discussed about the radon generation process also extend to the radon migration mechanisms. The analysis of the models developed and considered in the recent literature has revealed, other than a lack of standardization on parameters regulating the radon transport, some common simplifying assumptions underlying the formulations adopted. These assumptions limit the field of application of the radon transport models commonly considered. Moreover, the assumptions adopted have been often not declared, and this has resulted in improper applications documented in literature.

Furthermore, most of the models currently considered to assess the indoor radon concentration neglect the contribution of the advective transport through the building materials: the radon is assumed to travel only down to an activity concentration gradient. These assumptions may lead to underestimate the resulting indoor radon concentration in some specific circumstances.

This chapter provides a systematic and comprehensive description of radon transport in porous media by considering several contributions of different authors. The resulting homogenization is crucial to understand, and extend, the applicability field of the different formulations available in literature, and it may result useful to develop formulations of radon transport applicable for specific scenarios. Finally, large reviews have been carried out to report the typical values of most of the parameters regulating or influencing the radon transport.

DOI: http://dx.doi.org/10.5772/intechopen.111710

The discussion carried out in this chapter may lead to a better understanding of the radon migration phenomenon. This, if considered together with the description of the radon generation process carried out in chapter 2, may lead to improve the design of both preventive and remediation strategies aiming to reduce the exposure to radon indoors.

Author details

Christian Di Carlo*, Andrea Maiorana and Francesco Bochicchio
National Center for Radiation Protection and Computational Physics, Italian National Institute of Health, Rome, Italy

*Address all correspondence to: christian.dicarlo@iss.it

References

[1] Chisté V, Bé MM. LNHB/CEA - Table de Radionucléides. Technical report; Gif-sur-Yvette, France: LNHB/CEA; 2007

[2] Martin EJ. Physics for Radiation Protection. third comp ed. Boschstr (Germany): Wiley-VCH Verlag GmbH & Co. KGaA; 2013

[3] UNSCEAR. Sources and effects of ionizing radiation. In: United Nations Scientific Committee on the Effects of Atomic Radiation (UNSCEAR) 2008 Report. report to the General Assembly, with Scientific Annexes. New York, United States. 2008

[4] United States Environmental Protection Agency, editor. EPA Assessment of Risks from Radon in Homes. Washington DC: Office of Radiation and Indoor Air United States Environmental Protection Agency; 2003

[5] WHO. Handbook on Indoor Radon: A Public Health Perspective. Vol. 67. Geneva, Switzerland: W.H. Organization; 2010

[6] IARC. Radiation: A review of human carcinogens. In: W.H. Organization, editor. IARC Monograph on the Evaluation of Carcinogenic Risks to Humans. Lyon, France: International Agency for Research on Cancer; 2012

[7] Darby S et al. Radon in homes and risk of lung cancer: Collaborative analysis of individual data from 13 European case-control studies. BMJ. 2005;**330**(7485):223

[8] Krewski D et al. Residential radon and risk of lung cancer: A combined analysis of 7 north American case-control studies. Epidemiology. 2005;**16**(2):137-145

[9] Lubin JH et al. Risk of lung cancer and residential radon in China: Pooled results of two studies. International Journal of Cancer. 2004;**109**(1):132-137

[10] Vienneau D et al. Effects of radon and UV exposure on skin cancer mortality in Switzerland. Environmental Health Perspectives. 2017;**125**(6):067009

[11] UNSCEAR. Sources and effects of ionizing radiation. In: United Nations Scientific Committee on the Effects of Atomic Radiation (UNSCEAR) 1993 Report: Report to the General Assembly, with Scientific Annexes. New York, United States: UNSCEAR; 1993

[12] Iida T et al. Continuous measurements of outdoor radon concentrations at various locations in East Asia. Environment International. 1996;**22**:139-147

[13] Oikawa S et al. A nationwide survey of outdoor radon concentration in Japan. Journal of Environmental Radioactivity. 2003;**65**(2):203-213

[14] Nero AV, Nazaroff WW. Characterising the source of radon indoors. Radiation Protection Dosimetry. 1984;**7**(1–4):23-39

[15] Tokonami S et al. Characteristics of radon and its progeny concentrations in air-conditioned office buildings in Tokyo. Radiation Protection Dosimetry. 2003;**106**(1):71-76

[16] Iimoto T. Time variation of the radon equilibrium factor in a reinforced concrete dwelling. Radiation Protection Dosimetry. 2000;**92**(4):319-321

[17] Hopke PK et al. Assessment of the exposure to and dose from radon decay products in normally occupied homes.

Environmental Science & Technology. 1995;**29**(5):1359-1364

[18] Chen J, Harley NH. A review of indoor and outdoor radon equilibrium factors-part I: 222Rn. Health Physics. 2018;**115**(4):490-499

[19] Prasad M et al. Variability of radon and thoron equilibrium factors in indoor environment of Garhwal Himalaya. Journal of Environmental Radioactivity. 2016;**151**(Pt 1):238-243

[20] Ramachandran TV, Ramu MCS. Variation of equilibrium factor F between radon and its short-lived decay products in an indoor atmosphere. Nuclear Geophysics. 1994;**8**(5):499-503

[21] Collignan B, Powaga E. Impact of ventilation systems and energy savings in a building on the mechanisms governing the indoor radon activity concentration. Journal of Environmental Radioactivity. 2019;**196**:268-273

[22] Chao CYH, Tung TCW, Burnett J. Influence of ventilation on indoor radon level. Building and Environment. 1997; **32**(6):527-534

[23] Demoury C et al. A statistical evaluation of the influence of housing characteristics and geogenic radon potential on indoor radon concentrations in France. Journal of Environmental Radioactivity. 2013;**126**:216-225

[24] Ippolito R, Remetti R. Radon entry models into buildings versus environmental parameters, building shape and types of foundation. WIT Transactions on Ecology and the Environment. 2019;**236**:309-316

[25] Bossew P et al. Development of a Geogenic radon Hazard index-concept, history, experiences. New York, United States: International Journal of Environmental Research and Public Health. 2020;**17**(11):4134

[26] UNSCEAR. Sources and effects of ionizing radiation. In: United Nations Scientific Committee on the Effects of Atomic Radiation (UNSCEAR) 1993 Report: Report to the General Assembly, with Scientific Annexes. New York, United States: UNSCEAR; 2000

[27] Arvela H. Residential Radon in Finland Sources, Variation, Modelling and Dose Comparisons. Helsinki, Finland: Helsinki University of Technology; 1995

[28] UNSCEAR. Sources and effects of ionizing radiation. In: United Nations Scientific Committee on the Effects of Atomic Radiation (UNSCEAR) 2006 Report: Report to the General Assembly, with Scientific Annexes. New York, United States: UNSCEAR; 2006

[29] Nazaroff WW, Nero AV. Radon and its Decay Products in Indoor Air. New York, United States: Wiley-interscience Publication; 1988

[30] Collignan B, Lorkowski C, Améon R. Development of a methodology to characterize radon entry in dwellings. Building and Environment. 2012;**57**: 176-183

[31] UNSCEAR. Sources and effects of ionizing radiation. In: United Nations Scientific Committee on the Effects of Atomic Radiation (UNSCEAR) 1988 Report: Report to the General Assembly, with Scientific Annexes. UNSCEAR; 1988

[32] Font L, Baixeras C. The RAGENA dynamic model of radon generation, entry and accumulation indoors. The Science of the Total Environment. 2003; **307**(1–3):55-69

[33] Kozak K et al. Correction factors for determination of annual average radon concentration in dwellings of Poland resulting from seasonal variability of indoor radon. Applied Radiation and Isotopes. 2011;**69**(10):1459-1465

[34] Prasad G et al. Estimation of radon diffusion coefficients in soil using an updated experimental system. The Review of Scientific Instruments. 2012; **83**(9):093503

[35] Bochicchio F et al. Annual average and seasonal variations of residential radon concentration for all the Italian regions. Radiation Measurements. 2005; **40**(2–6):686-694

[36] Sundal AV et al. Anomalously high radon concentrations in dwellings located on permeable glacial sediments. Journal of Radiological Protection. 2007; **27**(3):287-298

[37] Di Carlo C et al. Extreme reverse seasonal variations of indoor radon concentration and possible implications on some measurement protocols and remedial strategies. Environmental Pollution. 2023;**327**:121480

[38] Soil Science Society of America. Glossary of Soil Science Terms 2008. Madison, United States: ASA-CSSA-SSSA; 2008

[39] Ishimori Y et al. Measurement and Calculation of Radon Releases from NORM Residues, in Technical Report 474. Vienna: International Atomic Energy Agency (IAEA); 2013

[40] Tanner AB. Radon migration in the ground: A supplementary review. Proc. Natural Radiation Environment. 1980;**III**:5

[41] Sakoda A, Ishimori Y, Yamaoka K. A comprehensive review of radon emanation measurements for mineral, rock, soil, mill tailing and fly ash. Applied Radiation and Isotopes. 2011; **69**(10):1422-1435

[42] Cinelli, G., M. De Cort, and T. Tollefsen, European Atlas of Natural Radiation, Publication Office of the European Union, Editor. 2019, European Commission, Joint Research Centre: Luxembourg.

[43] Jackson JA, editor. Glossary of Geology. Fourth ed. Alexandria, Virginia: American Geological Institute; 1997

[44] dos Santos Júnior JA et al. Radioactive disequilibrium and dynamic of natural radionuclides in soils in the state of Pernambuco—Brazil. Radiation Protection Dosimetry. 2018;**182**(4): 448-458

[45] Greeman DJ, Rose AW, Jester WA. Form and behavior of radium, uranium, and thorium in Central Pennsylvania soils derived from dolomite. Geophysical Research Letters. 1990;**17**(6):833-836

[46] Jönsson G. Soil radon depth dependence. Radiation Measurements. 2001;**34**(1–6):415-418

[47] Kovach EM. Meteorological influences upon the radon-content of soil-gas. Transactions, American Geophysical Union. 1945;**26**(2):241-248

[48] Koarashi J, Amano H, Andoh M. Environmental Factors Affecting Radon Exhalation from a Sandy Soil. Tokyo (Japan): Japan Atomic Energy Research Institute; 2000

[49] Chen C, Thomas DM, Green RE. Modeling of radon transport in unsaturated soil. Journal of Geophysical Research: Solid Earth. 1995;**100**(B8): 15517-15525

[50] Al-Shereideh SA, Bataina BA, Ershaidat NM. Seasonal variations and depth dependence of soil radon concentration levels in different geological formations in Deir Abu-said district, Irbid—Jordan. Radiation Measurements. 2006;**41**(6):703-707

[51] Winkler R, Ruckerbauer F, Bunzl K. Radon concentration in soil gas: A comparison of the variability resulting from different methods, spatial heterogeneity and seasonal fluctuations. Science of the Total Environment. 2001; **272**(1–3):273-282

[52] Gruber V et al. The European map of the geogenic radon potential. Journal of Radiological Protection. 2013;**33**(1): 51-60

[53] Perrier F, Girault F, Bouquerel H. Effective radium-226 concentration in rocks, soils, plants and bones. Geological Society, London, Special Publications. 2016;**451**(1):113-129

[54] Font L. Radon generation, entry and accumultion. In: Grup de Fisica de les Radiacions. Barcelona (Spain): Universitat Autònoma de Barcelona; 1997. p. 189

[55] Ciotoli G et al. Geographically weighted regression and geostatistical techniques to construct the geogenic radon potential map of the Lazio region: A methodological proposal for the European atlas of natural radiation. Journal of Environmental Radioactivity. 2017;**166**(Pt 2):355-375

[56] Giustini F et al. Mapping the geogenic radon potential and radon risk by using empirical Bayesian kriging regression: A case study from a volcanic area of Central Italy. Science of the Total Environment. 2019;**661**:449-464

[57] Tollefsen T, De Cort M, Cinelli G, Gruber V, Bossew P. Uranium Concentration in Soil. European Commission, Joint Research Centre (JRC); 2016. Available from: http://data.europa.eu/89h/jrc-eanr-04_uranium-concentration-in-soil

[58] Michalik B, de With G, Schroeyers W. Measurement of radioactivity in building materials – Problems encountered caused by possible disequilibrium in natural decay series. Construction and Building Materials. 2018;**168**:995-1002

[59] Ivanovich M. Uranium series disequilibrium: Concepts and applications. Radiochimica Acta. 1994; **64**(2):81-94

[60] Kekelidze N et al. Radioactivity of soils in Mtskheta-Mtianeti region (Georgia). Annals of Agrarian Science. 2017;**15**(3):304-311

[61] Phong Thu HN, Van Thang N, Hao LC. The effects of some soil characteristics on radon emanation and diffusion. Journal of Environmental Radioactivity. 2020;**216**:106189

[62] Bossew P. The radon emanation power of building materials, soils and rocks. Applied Radiation and Isotopes. 2003;**59**(5–6):389-392

[63] Krupp K, Baskaran M, Brownlee SJ. Radon emanation coefficients of several minerals: How they vary with physical and mineralogical properties. American Mineralogist. 2017;**102**(7):1375-1383

[64] Zhang W, Zhang Y, Sun Q. Analyses of influencing factors for radon emanation and exhalation in soil. Water, Air, & Soil Pollution. 2019;**230**(1):16

[65] Strong KP, Levins DM. Effect of moisture content on radon emanation from uranium ore and tailings. Health Physics. 1982;**42**(1):27-32

[66] Breitner D et al. Effect of moisture content on emanation at different grain size fractions - a pilot study on granitic esker sand sample. Journal of Environmental Radioactivity. 2010; **101**(11):1002-1006

[67] Sun Q, Zhang W, Qian H. Effects of high temperature thermal treatment on the physical properties of clay. Environmental Earth Sciences. 2016; **75**(7):610

[68] Han J et al. Experimental study on thermophysical properties of clay after high temperature. Applied Thermal Engineering. 2017;**111**:847-854

[69] Stranden E, Kolstad AK, Lind B. The influence of moisture and temperature on radon exhalation. Radiation Protection Dosimetry. 1984;**7**(1–4):55-58

[70] Tuccimei P et al. Radon exhalation rates of building materials: Experimental, analytical protocol and classification criteria. In: Cornejo DN, Haro JL, editors. Building Materials: Properties, Performance and Applications. Hauppauge, United States: Nova Science Publishers; 2009. pp. 259-274

[71] Iskandar D, Yamazawa H, Iida T. Quantification of the dependency of radon emanation power on soil temperature. Applied Radiation and Isotopes. 2004;**60**(6):971-973

[72] Maraziotis EA. Effects of Intraparticle porosity on the radon emanation coefficient. Environmental Science & Technology. 1996;**30**(8): 2441-2448

[73] Sakoda A et al. Experimental and modeling studies of grain size and moisture content effects on radon emanation. Radiation Measurements. 2010;**45**(2):204-210

[74] Abdel-Razek YA et al. Radon emanation coefficients of some Egyptian rocks. Journal of Radiation and Nuclear Applications. 2019;**4**(1):59-63

[75] Eckertova T, Bohm R, Holy K. Study of the influence of soil moisture and grain size on radon emanation from soil using an advanced multigrain model. Radiation Protection Dosimetry. 2022; **198**(9–11):778-784

[76] Ziegler JF, Biersack JP. The Stopping and Range of Ions in Matter. Treatise on Heavy-Ion Science. US: Springer; 1985. pp. 93-129

[77] de Martino S, Sabbarese C, Monetti G. Radon emanation and exhalation rates from soils measured with an electrostatic collector. Applied Radiation and Isotopes. 1998;**49**(4):407-413

[78] Hultquist B. Studies on naturally occurring ionizing radiations with special reference to radiation doses in swedish houses of various types. Kungl. Svenska Vetenskapsakademiens Handlingar. [PhD Thesis]. Almqvist & Wiksell: Stockholm College; 1956

[79] Kolb W, Schmier H. Building material induced radiation exposure of the population. Environment International. 1978;**1**(1–2):69-71

[80] Sorantin H, Steger F. Natural radioactivity of building materials in Austria. Radiation Protection Dosimetry. 1984;**7**(1–4):59-61

[81] Krisiuk EM et al. A Study of Radioactivity in Building Materials. Leningrad: Research Institute for Radiation Hygiene; 1971

[82] Sciocchetti G et al. Results of a survey on radioactivity of building

materials in Italy. Health Physics. 1983; **45**(2):385-388

[83] UNSCEAR. Sources and effects of ionizing radiation. In: United Nations Scientific Committee on the Effects of Atomic Radiation (UNSCEAR) 1977 Report: Report to the General Assembly, with Scientific Annexes. New York, United States: UNSCEAR; 1977

[84] Sabbarese C, Ambrosino F, D'Onofrio A. Development of radon transport model in different types of dwellings to assess indoor activity concentration. Journal of Environmental Radioactivity. 2021;**227**:106501

[85] Barros-Dios JM et al. Factors underlying residential radon concentration: Results from Galicia, Spain. Environmental Research. 2007; **103**(2):185-190

[86] Denman AR et al. Health implications of radon distribution in living rooms and bedrooms in U.K. dwellings–a case study in Northamptonshire. Environment International. 2007;**33**(8):999-1011

[87] Bochicchio F et al. Results of the National Survey on radon indoors in all the 21 Italian regions. In: Radon in the Living Environment. Athens; 1999

[88] Man CK, Yeung HS. Modeling and measuring the indoor radon concentrations in high-rise buildings in Hong Kong. Applied Radiation and Isotopes. 1999;**50**(6):1131-1135

[89] Shaikh AN, Ramachandran TV, Vinod Kumar A. Monitoring and modelling of indoor radon concentrations in a multi-storey building at Mumbai, India. Journal of Environmental Radioactivity. 2003; **67**(1):15-26

[90] Yarmoshenko IV et al. MODELING and justification of indoor radon prevention and remediation measures in multi-story apartment buildings. Results in Engineering. 2022;**16**:100754

[91] McGrath JA et al. Factors influencing radon concentration during energy retrofitting in domestic buildings: A computational evaluation. Building and Environment. 2021;**194**: 107712

[92] Frutos B et al. Inner wall filler as a singular and significant source of indoor radon pollution in heritage buildings: An exhalation method-based approach. Building and Environment. 2021;**201**: 108005

[93] Trevisi R et al. Updated database on natural radioactivity in building materials in Europe. Journal of Environmental Radioactivity. 2018;**187**:90-105

[94] Leonardi F et al. A study on natural radioactivity and radon exhalation rate in building materials containing norm residues: Preliminary results. Construction and Building Materials. 2018;**173**:172-179

[95] Labrincha J et al. From NORM by-products to building materials. In: Schroeyers W, editor. Naturally Occurring Radioactive Materials in Construction. Hasselt, Belgium: Woodhead Publishing; 2017. pp. 183-252

[96] Nuccetelli C, Leonardi F, Trevisi R. Building material radon emanation and exhalation rate: Need of a shared measurement protocol from the european database analysis. Journal of Environ-mental Radioactivity. 2020;**225**: 106438

[97] Garver E, Baskaran M. Effects of heating on the emanation rates of radon-222 from a suite of natural minerals. Applied Radiation and Isotopes. 2004; **61**(6):1477-1485

[98] Barton TP, Ziemer PL. The effects of particle size and moisture content on the emanation of Rn from coal ash. Health Physics. 1986;**50**(5):581-588

[99] Hassan NM et al. The effect of water content on the radon emanation coefficient for some building materials used in Japan. Radiation Measurements. 2011;**46**(2):232-237

[100] Somlai J et al. Connection between radon emanation and some structural properties of coal-slag as building material. Radiation Measurements. 2008;**43**(1):72-76

[101] Gijbels K et al. The influence of porosity on radon emanation in alkali-activated mortars containing high volume bauxite residue. Construction and Building Materials. 2020;**230**: 116982

[102] Pereira A et al. Estimation of the radon production rate in granite rocks and evaluation of the implications for geogenic radon potential maps: A case study in Central Portugal. Journal of Environmental Radioactivity. 2017;**166** (Pt 2):270-277

[103] Misdaq MA, Amghar A. Radon and thoron emanation from various marble materials: Impact on the workers. Radiation Measurements. 2005;**39**(4): 421-430

[104] Jobbagy V et al. Dependence of radon emanation of red mud bauxite processing wastes on heat treatment. Journal of Hazardous Materials. 2009; **172**(2–3):1258-1263

[105] Kuzmanovic P et al. The influence of building material structure on radon emanation. Journal of Radiological Protection. 2022;**42**(4): 041508

[106] Semkow TM, Parekh PP. The role of radium distribution and porosity in radon emanation from solids. Geophysical Research Letters. 1990; **17**(6):837-840

[107] Sas Z et al. Influencing effect of heat-treatment on radon emanation and exhalation characteristic of red mud. Journal of Environmental Radioactivity. 2015;**148**:27-32

[108] Li LG et al. Pervious concrete: Effects of porosity on permeability and strength. Magazine of Concrete Research. 2021;**73**(2):69-79

[109] Hall C, Hamilton A. Porosity–density relations in stone and brick materials. Materials and Structures. 2013;**48**(5):1265-1271

[110] Wedekind W et al. Weathering of volcanic tuff rocks caused by moisture expansion. Environmental Earth Sciences. 2012;**69**(4):1203-1224

[111] Raviv M et al. The effect of hydraulic characteristics of volcanic materials on yield of roses grown in soilless culture. Journal of the American Society for Horticultural Science. 1999; **124**(2):205-209

[112] Yan Z et al. Pore structure characterization of ten typical rocks in China. Electronic Journal of Geotechnical Engineering. 2015;**20**(2):479-494

[113] Straube JF. Moisture Control and Enclosure Wall Systems. Ontario, Canada: University of Waterloo; 1998

[114] Marshall TJ, Holmes JW, Rose CW. Soil Physics. Cambridge, England: Cambridge University Press; 1996

[115] Winslow DN. The pore size distribution of Portland cement paste. In: Station JHRPEE, editor. Fundamental

Studies in Portland Cement Concrete, Phase I. Lafayette, Indiana: Purdue University; 1968

[116] Ordóñez S, Fort R, Garcia del Cura MA. Pore size distribution and the durability of a porous limestone. Quarterly Journal of Engineering Geology and Hydrogeology. 1997;**30**(3): 221-230

[117] Kashif M et al. Pore size distribution, their geometry and connectivity in deeply buried Paleogene Es1 sandstone reservoir, Nanpu sag. East China. Petroleum Science. 2019;**16**(5):981-1000

[118] Stoulos S, Manolopoulou M, Papastefanou C. Measurement of radon emanation factor from granular samples: Effects of additives in cement. Applied Radiation and Isotopes. 2004;**60**(1): 49-54

[119] Moreno V et al. Radon levels in groundwaters and natural radioactivity in soils of the volcanic region of La Garrotxa, Spain. Journal of Environmental Radioactivity. 2014;**128**:1-8

[120] Fonollosa E et al. Radon in spring waters in the south of Catalonia. Journal of Environmental Radioactivity. 2016; **151**(Pt 1):275-281

[121] Auvinen A et al. Radon and other natural radionuclides in drinking water and risk of stomach cancer: A case-cohort study in Finland. International Journal of Cancer. 2005;**114**(1):109-113

[122] WHO. In: Press W, editor. Guidelines for Drinking-Water Quality. fourth ed. Geneva, Switzerland: WHO; 2017

[123] ICRP. ICRP Publication 137. Occupational Intakes of Radionuclides: Part 3. Ottawa, Canada: Annals of the ICRP; 2017

[124] Chen J. A discussion on issues with radon in drinking water. Radiation Protection Dosimetry. 2019;**185**(4): 526-531

[125] Morris BL, Lawrence ARL, Chilton PJC, Adams B, Calow RC, Klinck BA. Groundwater and its susceptibility to degradation: A global assessment of the problem and options for management. In: Early Warning and Assessment Report Series. RS. 03-3. Nairobi, Kenya: United Nations Environment Programme; 2003

[126] Chapman D, World Health Organization and UNESCO and United Nations Environment Programme. Water Quality Assessments: A Guide to the use of Biota, Sediments and Water in Environmental Monitoring. 2nd ed. Cambridge, England: E&FN Spon; 1996

[127] Phok R. Groundwater: Groundwater Resources, Challenges and Treatments Processes. Norway: University of Life Science (NMBU); 2019

[128] Carrard N, Foster T, Willetts J. Groundwater as a source of drinking water in Southeast Asia and the Pacific: A multi-country review of current reliance and resource concerns. Water. 2019;**11**(8):1605

[129] WHO/UNICEF. Joint Monitoring Program (JMP) for Water Supply, Sanitation and Hygiene - Report Progress on Drinking Water, Sanitation and Hygiene: 2000-2017: Special Focus on Inequalities. WHO/UNICEF; 2019

[130] Jobbagy V et al. A brief overview on radon measurements in drinking water. Journal of Environmental Radioactivity. 2017;**173**:18-24

[131] Todorovic N et al. Public exposure to radon in drinking water in Serbia.

Applied Radiation and Isotopes. 2012; **70**(3):543-549

[132] Somlai K et al. 222Rn concentrations of water in the Balaton Highland and in the southern part of Hungary, and the assessment of the resulting dose. Radiation Measurements. 2007;**42**(3): 491-495

[133] Nikolopoulos D, Louizi A. Study of indoor radon and radon in drinking water in Greece and Cyprus: Implications to exposure and dose. Radiation Measurements. 2008;**43**(7): 1305-1314

[134] Wallner G, Wagner R, Katzlberger C. Natural radionuclides in Austrian mineral water and their sequential measurement by fast methods. Journal of Environmental Radioactivity. 2008;**99**(7):1090-1094

[135] Gruber V, Maringer FJ, Landstetter C. Radon and other natural radionuclides in drinking water in Austria: Measurement and assessment. Applied Radiation and Isotopes. 2009; **67**(5):913-917

[136] Beyermann M et al. Occurrence of natural radioactivity in public water supplies in Germany: (238)U, (234)U, (235)U, (228)RA, (226)RA, (222)RN, (210)PB, (210)PO and gross alpha activity concentrations. Radiation Protection Dosimetry. 2010;**141**(1):72-81

[137] Lopes I, Madruga MJ, Carvalho FP. Application of liquid scintillation counting techniques to gross alpha, gross beta, radon and radium measurement in portuguese waters. In: International Conference Held in Szczyrk, Poland on 17–21 May 2004. Vienna, Austria: IAEA; 2004

[138] Henshaw DL et al. Radon in domestic water supplies in the UK. Radiation Protection Dosimetry. 1993; **46**(4):285-289

[139] Dongre S et al. Estimation of inhalation and ingestion dose due to radon concentration in drinking water samples of Shankaraghatta Forest environment, Karnataka, India. Indian Journal of Science and Technology. 2023; **16**(5):367-376

[140] Vuckovic B et al. Radon in drinking water from alternative sources of water supply in the north of Kosovo. Radiation Protection Dosimetry. 2023;**199**(1): 44-51

[141] International Organization for Standardization. ISO 13164-3:2013 Water Quality — Radon-222 Part 3: Test Method Using Emanometry. Geneva, Switzerland. 2013

[142] Cosma C et al. Radon in water from Transylvania (Romania). Radiation Measurements. 2008;**43**(8):1423-1428

[143] Przylibski TA et al. (222)Rn and (226)Ra activity concentrations in groundwaters of southern Poland: New data and selected genetic relations. Journal of Radioanalytical and Nuclear Chemistry. 2014;**301**(3):757-764

[144] Salonen L. Natural radionuclides in groundwaters in Finland. Radiation Protection Dosimetry. 1988;**24**:163-166

[145] Trautmannsheimer M, Schindlmeier W, Hübel K. Radon exposure levels of the staff in the drinking water supply facilities in Bavaria, in Ger. International Congress Series. 2002;**1225**:81-86

[146] Pereira AJSC et al. Evaluation of groundwater quality based on radiological and hydrochemical data from two uraniferous regions of Western Iberia: Nisa (Portugal) and Ciudad

Rodrigo (Spain). Environmental Earth Sciences. 2014;**73**(6):2717-2731

[147] Gibbons D, Kalin R. A survey of Radon-222 in ground water from the Sherwood sandstone aquifer: Belfast and Newtownards, Northern Ireland. Groundwater Monitoring & Remediation. 1997;**17**(2):88-92

[148] Suresh S et al. Measurement of radon concentration in drinking water and natural radioactivity in soil and their radiological hazards. Journal of Radiation Research and Applied Sciences. 2020; **13**(1):12-26

[149] Di Carlo C et al. Radon concentration in self-bottled mineral spring waters as a possible public health issue. Scientific Reports. 2019; **9**(1):14252

[150] Skeppstrom K, Olofsson B. A prediction method for radon in groundwater using GIS and multivariate statistics. Science of the Total Environment. 2006;**367**(2–3):666-680

[151] Bourgoignie RR et al. On the Rn-222 and Ra-226 concentrations in water from the Pletrou source (vise). Ann Belg. Ver. Stralingsbescherm. 1982;**7**:5-16

[152] Kasic A et al. Radon measurements in well and spring water of the Tuzla area, Bosnia and Herzegovina. Arhiv za Higijenu Rada i Toksikologiju. 2016; **67**(4):332-339

[153] Desideri D et al. 222Rn determination in drinkable waters of a central eastern Italian area: Comparison between liquid scintillation and gamma-spectrometry. Journal of Radioanalytical and Nuclear Chemistry. 2005;**266**(2): 191-197

[154] NMWE. Bottled water: Key statistics. 2019. Available from: https://naturalmineralwaterseurope.org/statistics/ [Accessed: 2023].

[155] Mayer PW et al. Residential End Uses of Water. Denver, United States: AWWA Research Foundation and American Water Works Association; 1999

[156] Deoreo WB et al. Residential End Uses of Water. Denver, United States: AWWA Research Foundation and American Water Works Association; 2016

[157] Bendito P et al. Study on Water Efficiency Standard. Brussels, Belgium: European Commission (DG ENV); 2009

[158] Danske Vandværkers Forening and Danmarks Statistik, Communication to European Environment Agency (EEA). Copenhagen, Denmark: Danmarks Statistik. 1997.

[159] Statistiska Centralbyrån. Vattenräkenskaper- en pilotstudie om uttag, användning samt utsläpp, fysiska och monetära data. No. 6. Stockholm, Sweden: Statistics Sweden. 2000

[160] Environment Agency/Ofwat. DETR Indicators for Sustainable Development D07. Birmingham, England: Environment Agency/Ofwat; 2000

[161] Statistics Norway, Communication to European Environment Agency (EEA). 1981.

[162] Lauri E. Veden käyttö Suomessa. Vol. 305. Helsinki, Finland: Suomen ympäristökeskus; 1999

[163] RIVM. Milieucompendium 1999: A6.4 Huishoudelijk waterverbruik per hoofd van de bevolking. In: Technical Report. Bilthoven, Netherlands: RIVM; 1999

[164] Schleich J, Hillenbrand T. Determinants of residential water demand in Germany. In: Technical Report. Karlsruhe, Germany: Institute Systems and Innovation Research; 2007

[165] Da Costa V, Jobard E, Marquay J, Ollagnon M, Plat B, Radureau S. Public water and wastewater Services in France Economic, social and environmental data. In: Technical Report. 6th ed. Issy-les-Moulineaux, France: BIPE - Le Vivaldi; 2015

[166] ISTAT. Censimento delle acque per uso civile. In: Technical Report. Rome, Italy: ISTAT; 2017

[167] Emilia-Romagna R. La risorsa acqua e il comportamento dei cittadini in Emilia-Romagna. Bologna, Italy: Ervet Emilia-Romagna; 2013

[168] Boyle RW, LXXXII. The solubility of radium emanation. Application of Henry's law at low partial pressures. The London, Edinburgh, and Dublin Philosophical Magazine and Journal of Science. 2009;**22**(132):840-854

[169] Battino R, Clever HL. The solubility of gases in liquids. Chemical Reviews. 1965;**66**:395-463

[170] Weigel, F., Radon. Chemiker Zeitung, 1978. 102: p. 287-299

[171] Partridge JE, Horton TR, Sensintaffer EL. A study of radon-222 released from water during typical household activities. In: Technical Report. Montgomery, AL: U.S.E.P. Agency; 1979

[172] Gesell TF, Prichard HM. The contribution of radon in tap water to indoor radon concentrations. In: Natural Radiation Environment III. Houston, TX, USA; 1980

[173] Hess CT, Weiffenbach CV, Norton SA. Variations of airborne and waterborne Rn-222 in houses in Maine. Environment International. 1982;**8**(1–6): 59-66

[174] National Research Council. Risk Assessment of Radon in Drinking Water. Washington, D.C: National Academies Press; 1999

[175] Vinson DS, Campbell TR, Vengosh A. Radon transfer from groundwater used in showers to indoor air. Applied Geochemistry. 2008;**23**(9): 2676-2685

[176] Nazaroff WW et al. Potable water as a source of airborne 222Rn in U.S. dwellings: A review and assessment. Health Physics. 1987;**52**(3):281-295

[177] Minkin L. Is diffusion, thermodiffusion, or advection a primary mechanism of indoor radon entry? Radiation Protection Dosimetry. 2002; **102**(2):153-162

[178] Shen L, Chen Z. Critical review of the impact of tortuosity on diffusion. Chemical Engineering Science. 2007; **62**(14):3748-3755

[179] Dullien FAL. Porous Media: Fluid Transport and Fluid Structures. New York, United States: Academic Press Limited; 1979

[180] Nimmo J. Porosity and pore-size distribution. In: Encyclopedia of Soils in the Environment. London: Elsevier/ Academic Press; 2004. pp. 295-303

[181] Culot MV, Olson HG, Schiager KJ. Effective diffusion coefficient of radon in concrete, theory and method for field measurements. Health Physics. 1976; **30**(3):263-270

[182] Ullman WJ, Aller RC. Diffusion-coefficients in nearshore marine-

sediments. Limnology and Oceanography. 1982;**27**(3):552-556

[183] Buckingham E. Contributions to our Knowledge of the Aeration of Soils. Washington, D.C: B.O.S. U.S. Dept. of Agriculture; 1904

[184] Matyka M, Khalili A, Koza Z. Tortuosity-porosity relation in porous media flow. Physical Review. E, Statistical, Nonlinear, and Soft Matter Physics. 2008;**78**(2 Pt 2): 026306

[185] Jonassen N, McLaughlin JP. Exhalation of Radon-222 from Building Materials and Walls. CONF-780422–(Vol2). United States: Lowder, W.M. 1980.

[186] Wilkening M. Radon in the Environment, Ed. Socorro, United States: Elsevier; 1990

[187] Youngquist GR. SYMPOSIUM ON FLOW THROUGH POROUS MEDIA diffusion and flow of gases in porous solids. Industrial & Engineering Chemistry. 2002;**62**(8):52-63

[188] Knudsen M. Die Gesetze der Molekularströmung und der inneren Reibungsströmung der Gase durch Röhren. Annalen der Physik. 1909; **333**(1):75-130

[189] Andreucci D et al. Fick and Fokker–Planck diffusion law in inhomogeneous media. Journal of Statistical Physics. 2018;**174**(2):469-493

[190] Hirst W, Harrison GE. The diffusion of radon gas mixtures. Proceedings of the Royal Society A. 1939; **169**(939):473–586

[191] Nazaroff WW. Radon transport from soil to air. Reviews of Geophysics. 1992;**30**(2)

[192] Epstein N. On tortuosity and the tortuosity factor in flow and diffusion through porous media. Chemical Engineering Science. 1989;**44**(3):777-779

[193] Keller G, Hoffmann B, Feigenspan T. Radon permeability and radon exhalation of building materials. Science of the Total Environment. 2001; **272**(1–3):85-89

[194] Narula AK, Chauhan RP, Chakarvarti SK. Testing permeability of building materials for radon diffusion. Indian Journal of Pure and Applied Physics Procedia. 2010;**48**(7):505-507

[195] Narula AK et al. Calculation of radon diffusion coefficient and diffusion length for different building construction materials. Indian Journal of Physics. 2009;**83**(8):1171-1175

[196] Chauhan RP, Nain M, Kant K. Radon diffusion studies through some building materials: Effect of grain size. Radiation Measurements. 2008;**43**: S445-S448

[197] Rogers VC, Nielson KK. Correlations for predicting air permeabilities and 222Rn diffusion coefficients of soils. Health Physics. 1991;**61**(2):225-230

[198] Yamazawa H et al. Radon exhalation from a ground surface during a cold snow season. International Congress Series. 2005;**1276**:221-222

[199] Leivo V et al. Air pressure difference between indoor and outdoor or staircase in multi-family buildings with exhaust ventilation system in Finland. Energy Procedia. 2015;**78**:1218-1223

[200] WHO. Guidelines for Indoor Air Quality: Selected Pollutants, B.O. Copenhagen, Denmark: WHO

European Centre for Environment and Health - Regional Office for Europe; 2019. p. 484

[201] Chauhan RP, Kumar A. A comparative study of indoor radon contributed by diffusive and Advective transport through intact concrete. Physics Procedia. 2015;**80**:109-112

[202] Arnold LJ. A scale model study of the effects of meteorological, soil, and house parameters on soil gas pressures. Health Physics. 1990;**58**(5):559-573

[203] Whitaker S. Flow in porous media I: A theoretical derivation of Darcy's law. Transport in Porous Media. 1986;**1**(1): 3-25

[204] Darcy H. The public fountains of the city of Dijon. In: Experience and Application Principles to Follow and Formulas to be Used in the, I.G.O.B.a. HIGHWAYS. Paris: Kendall Hunt Publishing; 1856

[205] Wang L et al. Experimental investigation of flow characteristics in porous Media at low Reynolds Numbers (Re→0) under different constant hydraulic heads. Water. 2019;**11**(11):2317

[206] Chaudhary K et al. The role of eddies inside pores in the transition from Darcy to Forchheimer flows. Geophysical Research Letters. 2011;**38**(24):L24405

[207] Bear J. Dynamics of fluids in porous media. Soil Science. 1975;**120**(2):162-163

[208] Muskat M. The Flow of Homogeneous Fluids Through Porous Media. Vol. 46. No. 2. Ann Arbor, United States: McGraw-Hill Book Company. 1937

[209] Forchheimer P. Wasserbewegung durch Boden. Zeitschrift des Vereins Deutscher Ingenieure. 1901;**45**: 1781–1788

[210] Dupuit J. Études théoriques et pratiques sur le mouvement des eaux. Paris: Libraire des corps impériaux des ponts et chaussées et des mines; 1863

[211] Carrigy NB et al. Knudsen diffusivity and permeability of PEMFC microporous coated gas diffusion layers for different polytetrafluoroethylene loadings. Journal of the Electrochemical Society. 2012;**160**(2):F81-F89

[212] Sparks DL. The chemistry of saline and sodic soils. In: Sparks DL, editor. Environmental Soil Chemistry. 2nd ed. Burlington, United States: Academic Press; 2003. pp. 285-300. ISBN: 978-0-12-656446-4

[213] Chuvilin E, Grebenkin S, Sacleux M. Influence of moisture content on permeability of frozen and unfrozen soils. Earth's Cryosphere. 2016;**1**:66-72

[214] Miao T et al. A fractal analysis of permeability for fractured rocks. International Journal of Heat and Mass Transfer. 2015;**81**:75-80

[215] Scheidegger AE. The Physics of Flow through Porous Media. 3rd ed. Toronto, Canada: University of Toronto Press; 1974

[216] Tollefsen T, De cort M, Cinelli G, Gruber V, Bossew P. Soil Permeability. European Commission, Joint Research Centre (JRC). 2018. Available from: http://data.europa.eu/89h/jrc-eanr-11_soil-permeability

[217] Scott AG. Radon sources, radon ingress and models. Radiation Protection Dosimetry. 1994;**56**(1–4):145-149

[218] Quenard DA et al. Microstructure and transport properties of porous building materials. Materials and Structures. 1998;**31**(5):317-324

[219] Scott AG. Modelling radon sources and ingress: In: The 1993 International Radon Conference. Denver, Colorado: AARST; 1993

[220] Dean SW et al. Correlation between water vapor and air permeability of building materials: Experimental observations. Journal of ASTM International. 2011;**8**(3)

[221] Zhang MH, Gjorv OE. Permeability of high-strength lightweight concrete. ACI Materials Journal. 1991;**88**(5): 463-469

[222] Monteiro P. Concrete: Microstructure, Properties, and Materials. M.-H. Publishing; 2006

[223] Carcasses M, Ollivier JP. Transport by permeation. In: Marchand J, editor. Engineering and Transport Properties of the Interfacial Transition Zone in Cementitious Composites, R.P. S.A.R.L. 1999. pp. 149-156

[224] Bustos F et al. Reducing concrete permeability by using natural Pozzolans and reduced aggregate-to-Pasteratio. Journal of Civil Engineering and Management. 2015;**21**(2):165-176

[225] Renken KJ, Rosenberg T. Laboratory measurements of the transport of radon gas through concrete samples. Health Physics. 1995;**68**(6): 800-808

[226] Rogers VC, Nielson KK. Data and models for radon transport through concrete. In: International Symposium on Radon and Radon Reduction Technology. 1992. pp. VI-3

[227] Sakai Y. Correlations between air permeability coefficients and pore structure indicators of cementitious materials. Construction and Building Materials. 2019;**209**:541-547